Quarterly Essay

Quarterly Essay is published four times a year by Black Inc., an imprint of Schwartz Media Pty Ltd. Publisher: Morry Schwartz.

ISBN 978-1-86395-582-9 ISSN 1832-0953

Subscriptions – 1 year (4 issues): $59 within Australia incl. GST. Outside Australia $89.
2 years (8 issues): $105 within Australia incl. GST. Outside Australia $165.

Payment may be made by Mastercard or Visa, or by cheque made out to Schwartz Media. Payment includes postage and handling.

To subscribe, fill out and post the subscription card or form inside this issue, or subscribe online:

www.quarterlyessay.com
subscribe@blackincbooks.com
Phone: 61 3 9486 0288

Correspondence should be addressed to:

The Editor, Quarterly Essay
37–39 Langridge Street
Collingwood VIC 3066 Australia
Phone: 61 3 9486 0288 / Fax: 61 3 9486 0244
Email: quarterlyessay@blackincbooks.com

Editor: Chris Feik. Management: Sophy Williams, Jess Tran. Publicity: Elisabeth Young. Design: Guy Mirabella. Assistant Editor/Production Coordinator: Nikola Lusk. Typesetting: Duncan Blachford.

AFTER THE FUTURE

Australia's New Extinction Crisis

Tim Flannery

> When it suits them, men may take control and play fine tricks and
> hustle Nature. Yet we may believe that Australia, quietly and imper-
> ceptibly ... is experimenting on the men ... She will be satisfied at
> long last, and when she is satisfied an Australian nation will in
> truth exist.
>
> —Sir Keith Hancock, *Australia* (1930)

This essay is an investigation into Australia's efforts to protect its endangered species from extinction. It focuses particularly on the effectiveness
of the federal legislation dealing with species officially recognised as being
under threat, but it also takes a broader view. How effective, for example,
have been state and federal efforts to preserve biodiversity by setting aside
national parks and nature reserves? Why are species still becoming extinct,
even though tens of millions of dollars are being spent to protect nature?
And what more needs to be done to prevent extinctions?

As I researched these issues, I grew increasingly dismayed at how haphaz-
ard and generally ineffectual our efforts at preventing extinctions have been.

In the twenty years since federal legislation was enacted, just one vertebrate species has increased in number sufficiently to be taken off the threatened list: the saltwater crocodile. Dismayingly, I also discovered that many conservative state governments are rolling back protections for nature, and that the worst are using aspects of our natural heritage as political bargaining chips. Yet some organisations and initiatives are making progress in protecting species – even bringing some back from the brink of extinction. Using them as models, I outline how private–public partnerships could conserve Australia's biodiversity effectively, and at a modest cost.

Australia is not alone in experiencing an extinction crisis. Many of our regional neighbours are in danger of losing their most distinctive species. I believe that Australian expertise could play a leading role in biodiversity protection regionally, and that a federal fund should be established to facilitate this.

Writing this essay has made me look at my society anew. Eighteen years ago I wrote *The Future Eaters*, which raised some of the issues discussed here. How much progress has been made towards sustainability since then? To answer that we need to revisit this essay's epigraph. When I first read Sir Keith Hancock's words, they seemed to leap from the page and sear themselves on my mind. Was ever a national narrative so perfectly distilled? Hancock's words capture the trajectory of all settler societies since the dawn of civilisation, but are particularly apt for Australia, where the mismatch between land and people was so profound, and the experience is still so raw.

I've spent my life in a country which looks upon its fine tricks and hustles of nature as some of its greatest achievements. We crow about them on the front pages of our newspapers, and look upon some as inexhaustible sources of wealth. But in reality we Australians are mere squatters in our own country. That is all we ever can be until Australia has completed its experiment. Moreover, our tenure in this land is limited not by some governor's pleasure, but by the rate at which we destroy its natural riches, including its species.

The great majority of Australia's plants and animals are found nowhere else on earth. Many are the result of 45 million years of separate evolution, for that is how long Australia has existed as an island continent. As a result, many Australian species are precious repositories of unique genes and evolutionary strategies, living in unique ecosystems. They are important not just in and of themselves, but because they provide Australians with the best means we have of engaging nature and listening to our land.

By learning about our homeland and adjusting our beliefs, values and practices, we can achieve great things. Indeed, over the past half-century, significant progress has been made in both caring for the Australian environment and in placing our culture on a more sustainable path. Yet in recent years things have begun to go backwards, as the concept of practical, measurable environmental protection has been widely neglected – even abandoned in some instances. I believe that two things – a lack of awareness of the severity of Australia's environmental problems, and the increasingly divisive, ideologically driven nature of our politics – are primarily responsible for this.

Although Australians profess to love their wildlife, there is an ever-growing sense among many of our politicians and business leaders that the natural world is something to be traded off – just another item in a ledger, or a criterion to be partially satisfied. This was highlighted to me recently in Kalgoorlie. I was talking to a group of school kids, one of whom asked me why trees were being knocked down to make way for new mines. When I responded that it was happening because our society values money more than Australia's natural habitats, an employee of a major mining company objected that this was not true. So I asked him what he would do if he discovered Uluru was rich in gold ore. After a moment's thought he replied: "Come up from underneath." In other words, hollow out the country: compromise its natural treasures. On a scale far beyond mere mining, that's what we're doing today.

In aspiring, as they increasingly seem to do, to little more than the accumulation of wealth, some Australians have cultivated an apparently benign indifference to the natural world. Moreover, among some on the right of politics there's a growing hostility towards anything "environmental," which extends even to the science that supports wise management. As a result some Australians have begun to shoot the messenger, by cultivating a deep hostility towards all scientific expertise. Although this is most evident with respect to climate change, it affects all aspects of environmental and even medical science (think of vaccines) and is occurring just as a new wave of animal and plant extinctions is gathering pace.

Scientific research must set the compass for us when it comes to preventing extinctions. Hence I will focus in this essay on the science of biodiversity conservation – specifically on the conservation of species – and the political and social changes that must occur if we are to preserve our unique plants and animals. Many ideologies travel under the banner of nature conservation nowadays, including animal rights, landscape preservation and even resource management. All are arguably important in their own right, but none should get in the way of protecting species.

That is a rather unfashionable view at present. Many scientists and land managers prefer to focus on ecosystem protection rather than the fate of individual species, and this has led them to give priority to setting aside representative samples of each of Australia's ecosystem types in reserves and national parks. Of course this is important work, but I will argue that in and of itself it will not result in biodiversity protection. Instead, experience shows that unless such areas are carefully managed, the outcome for biodiversity is likely to be very poor indeed. That's why I believe that species, as well as ecosystems and landscapes, must once more become an important focus of our conservation efforts.

By way of illustrating how scientific research can guide conservation, and in order to show how science works, I will discuss in detail an interpretation of Australian prehistory published in my book *The Future Eaters*. It put forward the hypothesis that the first humans to colonise the continent swiftly hunted its large animals to extinction, and that this altered Australia's vegetation, nutrient cycling, fire frequency and intensity, and even climate – changes that have great relevance for land management today. Some palaeontologists and experts in dating technologies set about testing the hypothesis almost as soon as it was published. As a result spectacular progress has been made in understanding Australia's prehistory. But it took until 2012 for a truly rigorous test to emerge. The key was an elegant – indeed beautiful – piece of science which I'll explain in detail later. It does not prove the Future Eaters hypothesis correct – for, contrary to popular opinion, it's impossible to prove anything in science. But it does represent a major step forward, and was a strong factor in prompting me to write this essay at this time.

Most of Australia's biodiversity consists of invertebrates such as insects and spiders (which make up 97 per cent of all animal species), and plants. Yet here I'll be focusing on the fate of vertebrates such as wallabies and bandicoots. This is in part because large creatures such as mammals play a disproportionately important role in seed and spore dispersal and nutrient recycling, which are vital "ecosystem services." Nor will I write much about threats such as mining and agriculture, simply because, while agriculture was an important threat in the past, and mining can have a locally catastrophic effect, today there are greater threats.

THE EXTINCTION PROBLEM

In late August 2009 a tiny, solitary bat fluttered about in the rainforest near Australia's infamous Christmas Island detention camp. We don't know precisely what happened to it. Perhaps it landed on a leaf at dawn after a night feeding on moths and mosquitoes, and was torn to pieces by invasive fire ants; perhaps it succumbed to a mounting toxic burden placed on its tiny body by insecticide spraying. Or maybe it was simply worn out with age and ceaseless activity, and died quietly in its tree-hollow. But there is one important thing we do know: it was the very last Christmas Island pipistrelle (*Pipistrellus murrayi*) on earth. With its passing, an entire species winked out of existence.

Two decades earlier the island's population of pipistrelles had been healthy. A few scientists had watched the species' decline with concern, until, after the million or more years that it had played a part in keeping the ecological balance of the island, they could see that without action its demise was imminent. They had done their best to warn the federal government about the looming catastrophe, but they might as well have been talking to a brick wall. The bureaucrats and politicians prevaricated for three years, until it was too late. While Australians argued about the fate of the asylum seekers who shared the pipistrelle's home, nothing effective was done to help the bats. Indeed, except for those few watching scientists, neither Australia's press nor public seemed to give a thought to the passing of the species, nor what it might mean for Christmas Island or our relationship with our country.

The pipistrelle's extinction was almost unbearably painful for me. In an attempt to avert it I had met with Peter Garrett, then the environment minister, and warned him of the impending loss. I had also brought offers of assistance and expertise from the Australian Mammal Society to his attention. The society was confident that the species could be saved – at a cost of perhaps only a few hundred thousand dollars. But Garrett was convinced by the orthodoxy that ecosystems rather than species should be the focus of the national conservation effort, and I got the message loud and clear that nothing would be done. Saving the bat wasn't an impossible mission: it's just that the government and the people of Australia – one of the richest countries on earth – decided it wasn't worth doing.

What really shook me about the episode was that it was the first extinction of a mammal to occur in Australia for sixty years – and therefore the first to occur in my lifetime. My original professional expertise lies in mammalogy and palaeontology, and before the pipistrelle's demise I had believed the worst of Australia's extinction crisis was behind us – that somehow my generation was wiser and more caring than earlier ones, and would not tolerate any more losses of Australia's unique mammals. It's now clear that those sixty years were just a lull in the storm, and that the pipistelle's demise marked the beginning of a new extinction wave.

Australia's first extinction wave started to gather pace almost as soon as the First Fleeters stepped ashore, and by the 1940s it had carried away 10 per cent of the continent's mammal species. No other class of organisms has suffered so grievously, and as a result mammals have become something of a yardstick by which we measure our long-term environmental impact. In 1791 a convict wrote about the white-footed rabbit rat, saying that it was a pest in the colony's food stores. The soft-furred, grey and white kitten-sized creature was arguably the most beautiful of Australia's seventy-odd native rodent species, yet it was destined to be one of the earliest victims of European settlement. Two hundred years ago it could be found in woodlands from near Brisbane to Adelaide, but the last record of it dates to the 1850s. Because foxes and rabbits had not begun to spread by this time, it is thought that a major factor contributing to its extinction was the end of Aboriginal fire management – a vital subject to which we shall return later in this essay.

The thylacine and the toolache wallaby were the largest creatures to succumb in the first extinction wave. Both had small populations and restricted distributions (Tasmania and the southeast of South Australia respectively), and are unique in being the only species that were hunted to extinction by Europeans. The thylacine was Australia's largest marsupial carnivore and, being wolf-like in appearance, it was persecuted by sheep farmers, the bounty on its head outlasting the creature itself. The beautiful toolache (pronounced "toolaitch") wallaby had the misfortune of being the fleetest member of the kangaroo family, and so was hunted for sport. These extinctions were, however, atypical: indeed, one of the most astonishing aspects of the first extinction wave was that its victims included what had been the most abundant and seemingly secure mammals in Australia.

Among the victims that once abounded were a dozen kitten- to hare-sized marsupials, mostly wallabies, rat-kangaroos and bandicoots, as well as nine species of native rodent. All of these species vanished between the 1840s and the 1930s, and all inhabited southern and central Australia. Strangely, many remained common until the moment of their vanishing. For example, according to the pioneering zoologist Frederic Wood Jones, in the early years of the twentieth century in Adelaide the rabbit-sized marsupials known locally as "tungoos" (brush-tailed bettongs and their relatives) were sold at nine pence per dozen for greyhounds to chase and kill. Yet just a few years later they were only a memory, with not so much as a single skin remaining in the state's museum.

The causes of these extraordinary extinctions are thought to have been varied. The cessation of Aboriginal burning doubtless had its effect, and until the 1930s bounties were paid by many state governments for the scalps of now-extinct creatures. But the depredations of foxes (which were spreading quickly by the early twentieth century) and feral cats, and the wholesale destruction of native vegetation by livestock and rabbits, must also have been important causes. While the causes are disputed, the effect of the first extinction wave is clear: it gutted the biodiversity of the drier parts of the continent, and very few native mammals larger than a rat and smaller than a kangaroo can be found on Australia's inland plains today. It's the absence of such species – the so-called critical-weight-range mammals (which weigh between 500 grams and 5 kilograms), which were once among the most abundant of creatures – that has led me to characterise the national parks of Australia's southern inland as "marsupial ghost towns."

The gathering second extinction wave is now mopping up the few surviving medium-sized mammals in Australia's south and inland. It's not difficult to predict which will be the next to become extinct, for, like the pipistrelle, their decline has been charted for years. There are 15 frogs, 16 reptiles, 44 birds, 35 mammals and 531 plants on Australia's endangered species list, and among those closest to the brink are three mammals: the central rock rat, the bridled nailtail wallaby and the numbat. All hang by a thread, and with a single vexed exception, which I will discuss at length later, next to nothing effective is being done to halt their slide into oblivion.

The most dismaying aspect of the second extinction wave is that it is emptying vast swathes of the continent that were untouched by the first wave. Australia's Top End and Kimberley were, until recently, a paradise for medium-sized mammals, among them a close relative of the white-footed rabbit rat. As will be explained later, the last two decades have seen this fauna all but exterminated in the Top End, even in our most valued and best resourced national parks.

Perhaps it is excusable that Australians are unaware of the extinctions currently occurring in distant places like Arnhem Land and other regions of our far north. But astonishingly, we also seem blind to the perils facing species much closer to home – for example, the sand flathead of Port Phillip Bay. A fish familiar to every Melburnian who has ever dangled a line, its population has declined by 97 per cent over the past decade. That means that just three fish survive for every 100 that were present in 2002. While the reasons for the decline are unclear and may be multiple, over-fishing is certainly a factor. Yet many recreational fishermen still angrily refuse to countenance the development of a system of marine reserves extensive enough to give the species a chance. I've seen tinnies lined up like taxis at an airport cab rank along the edge of the pathetically small Ricketts Point Marine Sanctuary (just 115 hectares), hoping to hook one of the fish it shelters. Because fishing is prohibited in the sanctuary, the place is a haven for amazing creatures, including the Port Jackson sharks that have been eliminated elsewhere in the bay. It is a reminder of what an abundant marine environment Port Phillip Bay once was, and could be again under good management.

Why should the extinction of Australian organisms concern us? I've had people tell me: "I don't give a stuff about cute furry animals. What have they ever done for me?" The answer, I think, is almost precisely the same as to the question of why human rights are important, even when they concern people we'll never meet. First and foremost, it is a matter of values. The demise of a bat may not weigh greatly in the balance of human wellbeing, but it speaks volumes about the human soul. Do we wish to be despoilers and executioners of the natural world? Or do we want our children to have the opportunity to enjoy a world as bountiful and diverse as the one our parents bequeathed to us?

As with human rights, extinctions beg the question of where we draw the line. If we can stand by as a species of bat is snuffed out, then why not other species as well? Can we really expect poor Indian villagers to heed our pleas to conserve the tigers that menace their livestock if we do nothing to prevent the extinction of Australian species? As with the question of torture, to open the door to the practice of extinction is to contemplate the horrific becoming routine.

The extinction of species also involves practical considerations. Australia only had around 277 species of land mammal at the time of European settlement, and around 10 per cent of those are already gone. In an ecosystem every species plays a role: small bats eat insects such as mosquitoes, which carry disease, and moths, whose larvae if left unchecked can damage mighty trees; rock rats carry seeds of economically and environmentally important plant species such as the quandong. Although it's often difficult to gauge the damage done to ecosystems by an extinction, especially in the short term, it is inarguable that ecosystems need a diversity of species to function efficiently.

Moreover, it is living things that make our earth habitable. It's been said so often that it has become mundane, but it is axiomatic that ecosystems manufacture or cleanse the air we breathe, the water we drink and the food we consume. As Paul Ehrlich says, breaking the links of an ecosystem through causing extinctions is like removing the rivets, one by one, from a jet aircraft. There's plenty of redundancy in the system, so the loss of a single species, or rivet, will not usually cause a catastrophe. But remove enough of them and there will be consequences. You can also think of extinctions as the economic equivalent of not training enough electricians or geologists. Eventually whole sectors of the economy will founder, until the cascade of job losses becomes overwhelming.

At the heart of this nation's efforts to save its endangered species is a register of subspecies, species and ecological communities that are threatened with extinction. By law, each entity included on the list should have a detailed recovery plan written for it, which when implemented should save it from extinction. These plans classify species on a sliding scale – from vulnerable (the category for the least endangered) to critically endangered or extinct. The federal legislation governing these plans states: "Recovery plans are binding on the Australian Government – once a recovery plan is in place, Australian Government agencies must act in accordance with that plan."

What a wonderful reassurance! It's a pity, then, that the system underpinning the promise is as rotten as Miss Havisham's wedding cake. By their fruit ye shall know them: since the legislation mandating action plans was enacted in 1992, only a single vertebrate species has become so abundant as to merit being taken off the threatened species register. But saltwater crocs are atypical of Australia's endangered species in that the threat they faced was simple: when the shooting for skins was stopped, the species recovered.

Why are we failing so abjectly in protecting our threatened species? The pitifully slow rate at which recovery plans are being drafted is one factor. In New South Wales, for example, in the last twenty years recovery plans have been completed for only around 10 per cent of all species listed as vulnerable to extinction. At that rate, it would take two centuries just to draft the recovery plans that are merely the first step to protect the state's threatened fauna!

Things get worse. In 2006 the federal government excused itself from the obligation to draft plans for species listed as vulnerable to extinction. As a result, if the Minister for the Environment decides for whatever reason not to draft a plan, then it simply isn't done. And even if a plan is completed, there's no guarantee that it will receive funding. Indeed there is no specific pool of funding dedicated to financing recovery plans. Instead, individuals interested in protecting endangered species must apply for grants wherever they can, and often the granting bodies put species protection low on their list of priorities. Most disconcertingly, there is no obligation on anyone to report on actions taken under the plans, so it is almost impossible to gain a detailed understanding of what works to protect species from extinction, and what does not.

A review by the World Wildlife Fund (WWF) of conservation plans written for members of the kangaroo family (of which there were fifty-six in Australia before European settlement, seven being already extinct) could identify only a single species whose status improved as a result of a plan: the burrowing bettong (*Bettongia lesueur*). This rabbit-sized creature had become extinct on the mainland around seventy years ago, but survived on some islands. Its status was changed from "endangered" to "vulnerable" because a population established on the mainland at Heirisson Prong in Shark Bay, Western Australia, had begun to breed. The population has now increased to around 450 individuals, which is a great outcome. But how much did we spend, overall, to achieve this single success? And why did it work when so much else has failed?

It is clear that we cannot assume that the road to recovery for endangered species will always be smooth. A close relative of the burrowing bettong, the woylie (*Bettongia penicillata*, one of the species known as "tungoos" in the Adelaide area), had actually been taken off the threatened species list in 1996 because of a similar recovery. But its population soon crashed again, and today it is listed as critically endangered.

Why are action plans so often failing to help species recover? The glacially slow development of the plans, along with the lack of obligation to fund and report back on them, are clearly major impediments. But there are other problems as well. Peter Cosier, who worked under Senator Robert Hill when he was Minister for the Environment in the late 1990s, says that most of the plans he saw did not describe how species might be saved. Instead, they often stated that more money was required for research before appropriate action could be taken. Michael Kennedy of the Humane Society International adds that the pot of money for environmental protection gets smaller and smaller every year. And of course there are no consequences to anyone for failure.

While researching this essay I could not locate an overview of the status of actions being taken under conservation plans – which is telling in itself. A plethora of organisations – federal, state and NGO – are involved, and the information on listed species is so scattered, partial and out of date that it's impossible to gain an overview. I came away from the attempt with a firm impression that, as things stand, species recovery plans will continue to accumulate in departmental filing cabinets, where they will be safely stored away, while the frogs, lizards, birds and mammals that grace our country slip inexorably into extinction.

The plight of the central rock rat (*Zyzomys pedunculatus*) offers a final infuriating example of how the current system is failing us. It's an ancient Australian rodent whose ancestors arrived on the continent over 4 million years ago. Sandy-coloured and with a carrot-like tail that thickens towards the base, it abounded a century ago in the rocky ranges of central Australia. Then it vanished, and for decades was thought to be extinct – until a colony was located by zoologists in 1995. Eventually populations were located at seventeen locations, and some individuals were taken into captivity, where they bred prolifically. Their reproductive success was to prove their downfall, however, for with nowhere to release the progeny, the breeding program was stopped. Tragically, the wild populations then vanished and the species was again presumed extinct. Finally, in 2010, just four individuals were located on the summit of Mount Sonder – which at 1380 metres is the fourth-highest mountain in the Northern Territory. During recent surveys only a single individual was seen there – though, almost miraculously, another population was located this year atop the 1389-metre-high Mount Giles.

I was so concerned about the imminent extinction of the central rock rat that I offered a donation of $15,000 to kick-start a program to save it by bringing some of the last surviving individuals into captivity. Despite this offer, and the fact that a most splendid recovery plan is being final-ised, absolutely nothing practical is being done to rescue the species from extinction. In fact, it is possible that the rodent is already extinct. If so, in what may be typical of the fate of species carried off by the second great extinction wave, it's as much a victim of apathy and ignorance as malice and greed.

Such is the depth of public ignorance about Australia's extinction crisis that most people are unaware that it is occurring, while those who do know of it commonly believe that our national parks and reserves are safe places for threatened species. In fact the second extinction wave is now in full swing, and it's emptying our national parks and wildlife reserves as ruthlessly as other landscapes. This is disturbing: national parks exist explicitly to conserve biodiversity, and their failure to do so is a failure both of government policy and of our collective will to protect our natural heritage. Paradoxically, biodiversity is sometimes flourishing more vibrantly on private land than in national parks, despite hundreds of millions of dollars being spent annually by our governments on reserved lands.

The problem lies not with the parks' staff, who are often dedicated and skilled at their work. Nor does it lie solely with budgets, although more funding rather than more cuts would always be welcome. Instead, the difficulties are at least threefold. First and foremost, the problem stems from the delusion that the simple act of proclaiming a national park or nature reserve will result in the protection of biodiversity. Parks must be proclaimed *and* effectively managed if biodiversity is to be protected. Secondly, the various government agencies responsible for biodiversity protection have allowed their scientific capacity to erode to the point where it's hard to be sure how many individuals of most endangered species survive; and thirdly, the attempt to save endangered species involves risks that bureaucracies are increasingly unwilling to take. The first duty of the bureaucrat these days seems to be to protect their minister from criticism: thus it often seems preferable to let a species die out quietly, seemingly a victim of natural change, than to institute a recovery program that carries a risk of failure, however small.

I hope the message is loud and clear. Australian politics, and the bureaucracy that supports it, is failing in one of its most fundamental obligations to future generations: the conservation of our natural heritage. Nor does that failure result entirely from the factors I have just identified. The times also suit cynical self-interest: cash-starved state governments, ever more desperate for income and political support, are rolling back even the inadequate protections that presently exist, and economic pressures are making it difficult for not-for-profit organisations that focus on nature protection to make ends meet.

What to do? As this saga of ignorance, folly and malice unfolds, it has become clear to me that those working outside government have a crucial role to play in conserving our biodiversity. Indeed, I believe that it is action by the private and not-for-profit sectors, working hand in hand with government, that holds the key to protecting our endangered species in a competent and affordable manner. But before I outline how this might be done, we Australians need to take a look at ourselves.

OLD OR NEW?

I commenced this essay with a quote from Sir Keith Hancock's book *Australia* because it set me thinking about the relationship Australians have with their country. Is it destructive and parasitical, or might the continent be "quietly and imperceptibly" experimenting on its people to create a nation of true Australians – one with a sustainable relationship with their land? Hancock's ruminations are now more than eighty years old, and it's a fair question as to whether we're any more "Australian" today than we were in 1930.

Mighty influences have been at play in this country over the past eighty years – influences that Hancock could scarcely have imagined. Today, more than one-quarter of Australians were born elsewhere, and with so many of us living in cities, our experience of the natural world is limited. Surely these trends have had an influence on our attitude towards it. Moreover, globalisation, the increasing polarisation of politics and the growing gap between rich and poor seem to have made us less secure and less caring.

Increasingly we live in a country where many people appear to have turned their backs on nature. Fuck the environment. Fuck action on climate change. Money – big money – that's the only thing that matters now, and everyone and everything holding back the flow of cash needs to get out of the way. Whether it's fertile farmland, marine sanctuaries, important nature reserves or even national parks – in Australia all are increasingly negotiable, so long as the money's right.

Is this attitude a recent phenomenon – a result of the resources boom, perhaps, or a reaction by the far right to the rise of the Greens? Or has it been a core aspect of our culture since settlement? Clearly, the first Europeans arrived in Australia ignorant of the conditions prevailing in their new home. And, of course, the same must have been true of the ancestors of the Aborigines when they arrived 45,000 years ago. Australia's "great experiment" with their descendants has produced one of the most environmentally aware cultures on earth. But the 224 years that have passed since the arrival of the First Fleet amount to just three human life-times. It's a fair question whether this second experiment will yield fruit before it's too late – before we destroy our biodiversity and irrevocably mar the beauty of our land.

I'm not the first biologist to fret over such things. In 1938, less than a decade after Hancock published his thoughts, the pioneering zoologist Alan "Jock" Marshall penned a jeremiad for Sydney's *Daily Telegraph* which, by way of a summing-up of the Australian character, provided a snapshot of what the great experiment had produced during its first century and a half. Marshall was already well known to Australian radio listeners as "Jock the Backyard Naturalist" for *The Argonauts Club*. He would go on to forge an illustrious career, founding Monash University's department of zoology and writing the 1966 classic and foundation text of Australian environmental conservation, *The Great Extermination: a Guide to Anglo Australian Cupidity, Wickedness & Waste*. It is typical of Marshall's style that his publisher found the book's full title too confronting to print on its cover. That directness, however, pales when compared with what he had to say in the *Daily Telegraph* in 1938.

In response to the *Telegraph* piece, the newspaper received 426 letters, 412 of which expressed "a personal dislike" of the author. The dislike evidently went rather deep, with Marshall reporting, "A friend whom I'd known since my first long pants refused to speak to me for a fortnight" after he read it, while a schoolteacher "abused me for 16 pages in iambic pentameters. I recall how cleverly she rhymed the word 'dastard' with one my publisher forbids me to mention."

Marshall's 1938 article was reprinted in 1942 in the extraordinary Think – Or Be Damned series. Published by Angus & Robertson, it was a sort of precursor to the Quarterly Essay. At a time when the Japanese were threatening imminent invasion, Marshall opened his contribution, titled *Australia Limited*, with an assessment of the country that might have left some wondering whether their home was worth fighting for:

> Australia ... is a third-rate country – a vast, parched lump of a
> continent ... It is doubtful if it could ever support more than 20 or
> 30 million people at the most.

This jarred violently with the young nation's view of itself. Fed on the opinions of "boosters" like Edwin Brady, whose 1918 book *Australia Unlimited* portrayed the country as a cornucopia, the popular view was that Australia had the potential to become a superpower – an antipodean United States supporting a population of hundreds of millions.

If Marshall's view of our continent was distasteful to the average Australian, his assessment of our politics and popular wisdom was all but despairing:

> The Australian attitude is that if a fellow still calls out "Good-day
> Jack" to you after you have elected him to parliament, then he is
> a good fellow, and necessarily a worthy parliamentarian. It mat-
> ters little his standing in the community, his intelligence or his
> learning.

Moreover, he added:

> No sane man would dare suggest that Sydney Harbour bridge was
> not the best in the world. I feel for our children, for by the time
> they grow up there'll probably be several better and bigger bridges
> ... But they'll still have to pay for the one we built!

Time has proved Marshall correct in so much: our population hovers within his predicted band, and of course there's still a toll on the Sydney Harbour Bridge – which is indeed no longer the largest on earth. Moreover, our politics remains instantly recognisable from his seventy-year-old characterisation. But to Marshall these were mere fripperies: our most defining – and appalling – characteristics, he felt, lay in our attitudes to education, money and the environment. "Who among us can give more than vague information about such great Australians as Grafton Elliot

Smith, Gilbert Murray, Edgeworth David, Parkes, Tilley, Mawson and the rest?" he moaned.

> About one in every five or six hundred has heard of Bertrand Russell … Money is the Australian's god … whatever wealthy Australians do with their money they use very little of it towards their country's advancement.

Marshall's Australia was a land of material affluence and intellectual destitution, one in which the *nouveau riche* – the Palmers, Tinklers and Rineharts of the day – contributed little or nothing to the betterment of their nation.

So what, as Marshall saw it, was the relationship between this distinctly unattractive society and its environment? The average Australian "knows nothing about the flora and fauna of his country and does nothing to help preserve it," he opined, going on to state that:

> It has always been strange to me that Australians, enamoured as they are of their "best country," should have slashed down most of its trees, eroded much of its pastures into deserts, left its cities without parks, and littered its bushland with tin cans and chocolate wrappers …

Here, I think, at last we can find evidence for a social change that may constitute proof of Australia "experimenting on the men." Looking at pictures of Australian cities and landscapes from the 1940s reveals a grim panorama: air pollution obscures the horizon, and there's barely a trace of greenery to interrupt the roads, tin roofs and forests of wire-strung telegraph poles that were the human habitat. In the decades since Marshall wrote, this has changed irrevocably. Moreover, a great deal has been done to protect Australia's forests, prevent soil erosion, create parks in our cities and deter litterbugs. All of this is evidence that the relationship between Australians and their environment has changed dramatically.

In the last half-century, in fact, care for the environment has become deeply ingrained in the national psyche. In Marshall's day there were a few groups, such as bird watchers and the Gould League, who sought to protect nature, but today there are hundreds, if not thousands, of such groups. From the Landcare movement to the myriad self-organising bands of bush regenerators and carers for local parks and gardens that have sprung up continent-wide, we Australians have become great participants in grassroots movements that seek to redefine our relationship with nature. And through such associations ordinary Australians have achieved extraordinary things, returning waterways to more natural and healthy states, controlling weed infestations, and monitoring and protecting native species as they return to habitats from which they've often been absent for decades.

Some groups have taken on more activist roles, and a few have begun to be concerned with endangered species. One such group, Environment East Gippsland, was formed by a group of volunteers dedicated to protecting the endangered species and old-growth forests of eastern Victoria from their own state government. In 2011 they took VicForests to court for failing to search for the endangered long-footed potoroo in old-growth forests that were scheduled to be logged – and they won. The court censured officialdom for neglecting its own regulations, which was an acute embarrassment to the state government. But what sort of world do we live in where individuals have to volunteer their time and energy to restrain a state government intent on destroying the environment that it is charged to protect?

Just fifty years ago both left and right were in furious agreement on the need for environmental conservation. Back then, as William Souder (Rachel Carson's most recent biographer) says:

> Conservation was a hopeful, noble and inherently nonpartisan cause, a quest for the betterment of the world. Its foundation was the conviction that we could protect and enjoy what nature provided – in perpetuity. From generation to generation, a well-tended earth would endure.

Recently, a disgruntled or self-interested few have began to attack any-thing they see as "green." Unfortunately, they are rapidly finding political allies, and as a result contempt by some on the far right of politics for all things environmental is growing. This is a new and disturbing trend that has taken deep root in Australian, American and Canadian politics. It is also profoundly anti-patriotic.

"Opinionists" of the extreme right are spearheading the initiative to divorce conservatives from conservation. In the United States, for example, they have worked tirelessly to destroy the Endangered Species Act. Here in Australia it's not so much endangered species as the concept of environ-mental stewardship that's the focus of their campaign. Chris Berg of the Institute of Public Affairs, commenting on the draft national curriculum in the Age on 8 July 2012, wrote:

> The suggestion we have a duty to be "stewards" of the environment comes straight from green political philosophy. It reduces humans to mere trustees of nature. This directly conflicts with the liberal belief that the Earth's bounty can be used for the benefit of humanity.

Does the idea that we should be stewards of the environment really come from some undefined "green" philosophy? "Stewardship" is commonly defined as responsible planning for the use and management of resources – which is surely universally acknowledged common sense rather than an ideological or political viewpoint. Is Berg ignorant of this, or is he trying to sell his political allies a shit sandwich so profoundly malodorous that it threatens to destroy the credibility of the right as a whole?

Far from "environmental stewardship" being a lefty philosophy, it has inspired some of the right's most prominent leaders, including Theodore Roosevelt, Ronald Reagan, Malcolm Fraser, Rupert Hamer and Garfield Barwick, to patriotic acts of enduring value. It was Fraser who in 1975 first used federal powers to prevent sand mining on Fraser Island, who proclaimed Kakadu a national park, and who ended whaling in Australia. Even groups caricatured as rednecks, such as some duck shooters, must be incensed at Berg's slur. It was, after all, Victorian duck shooters who in 1959 led the charge for environmental protection in their state when they enthusiastically agreed to impose an annual one-pound licence fee upon themselves, the proceeds of which were to be used to fund a wild-fowl management program – itself a fine example of stewardship of the environment.

Neither was the Victorian government of the time, though of a distinctly right-wing persuasion, niggardly in supporting environmental steward-ship. Before 1959 no land whatsoever had been set aside in Victoria explic-itly for wildlife conservation, yet by 1966, with the help of the duck shooters levy, 100,000 acres had been reserved for the protection of water fowl, seabirds and rock wallabies among other species, and a string of research stations covering many different habitats was being planned. Moreover, as Jock Marshall stated in *The Great Extermination*:

> I have said many harsh things about politicians in these pages. It is now my pleasure to say that the Victorian conservation programme has had at all times the support of the Victorian Premier, Mr Henry Bolte, and the Chief Secretary, Mr A.G. Rylah.

So how is it that today in the United States, Canada and most particularly Australia, some on the far right can deride the concepts of conservation and environmental stewardship? According to Souder, the alienation of the right from environmental concerns began with Rachel Carson's challenge to the chemical industry in her 1962 classic *Silent Spring*:

> There is no objective reason why environmentalism should be the exclusive province of any one political party or ideology – other than the history of the environmental movement beginning with *Silent Spring*. The labels for Carson rained down on her like fallout: subversive, anti-business, Communist sympathiser, health nut, pacifist, and, of course, the coded insult "spinster." The attack … came from the chemical companies, agricultural interests, and the

allies of both in government – the self-protective enclaves within what President Eisenhower had called the "military-industrial complex." Their fierce opposition to *Silent Spring* put Rachel Carson and everything she believed about the environment firmly on the left end of the political spectrum. And so two things – environmentalism and its adherents – were defined once and forever.

The impression that protecting the environment is primarily an issue for the left, and that industry acts as a bloc to prevent environmental protection, is now widespread. But it is entirely wrong. The fact that increased environmental awareness has seen costs or regulations imposed upon polluters, who complain – often loudly – about it, reinforces this preconception. But industry does *not* act as a bloc in this regard. One reason for this is that for every business that is made to clean up its act, there's another that is set to profit from the change. Industry is no monolithic enemy of environmental action: indeed, as the CEO of one multinational company recently said to me, environmental protection is no longer about window-dressing – it's about how you run the shop.

This is not to say that massively powerful vested interests do not at times act in concert to frustrate environmental protection. Industry peak bodies will on occasion oppose it if they feel that their membership might see such action as justifying the peak body's existence; and sometimes individual business leaders will join forces to thwart environmental stewardship. Some coal barons, for example, feel threatened by efforts to combat climate change, and have organised an effective disinformation campaign, particularly in the United States, to frustrate action on the issue.

Increasingly, those who act to frustrate environmental protection – whether it be by opposing the creation of marine reserves or opposing action on climate change – also include a group of disempowered, mostly older males who feel enraged that the world around them is changing, and so leaving them and their achievements behind. I've seen this sort of thing before, in Papua New Guinea, where social change has been so swift that village elders find their authority and status threatened by a rising younger generation. Sometimes you can find three generations of men in the one hut: Granddad often sits in the corner, telling anyone who'll listen about the good old days, when you made your name through rape and pillage. Dad will affect to be horrified by this, and start to talk about how joining the church is the best way forward. After all, that's how he stole the march on the big men and witchdoctors of his father's generation. And all the while the grandson will be looking out the door, playing with his mobile phone and dreaming of making his way in a far larger world. These old men deserve our sympathy. In earlier times they'd be enjoying an esteemed and respected old age. Instead, the young increasingly see their triumphs as criminal acts. In this they remind me of many younger Australians, who see big polluters and the older generations as criminally negligent in their attitude to the environment.

In Australia, those who feel that they are being disenfranchised include princes of the Catholic Church, such as Cardinal George Pell, who see their grip on the populace loosening by the year. Some retired scientists and engineers, who never adapted to the computer age and who have joined the ranks of the climate denialists, can also be placed in this category. Then there are the leaders of the generally disgruntled, such as Alan Jones. It's important that those in the environment movement understand that these grumpy old men do not represent the business community, nor even the right of politics. Indeed, as can be seen from the likes of Robert Purvis, Malcolm Turnbull and Martin Copley, both business and the political right, broadly defined, contain at least as many allies as enemies in the fight for environmental protection.

There are further complexities to the growing political divide over the environment. The rise of the Greens and their alliance with Labor guarantees that some conservatives will have a knee-jerk reaction to environmental issues. And the fact that the Greens are not exclusively focused on environmental outcomes, but also have policies on animal and civil rights – which the public often confuses with environmental protection

– complicates matters still further. The fact is that animal-rights issues, such as opposition to the culling of feral species, can sometimes get in the way of environmental stewardship, and concerns about animal suffering need to be treated separately.

Party and ideological allegiances here work to prevent recognition of common interests and values. The far right consists of an amalgam of two largely incompatible philosophical groups – extreme free-market libertarians and deep conservatives – who remain allied solely through shared hatred of everything to their left. Libertarians tend to resent government and the constraints it places on human freedom, and consequently many environmental regulations are anathema to them. Conservatives, on the other hand, are quite different. They would like to preserve the status quo and, as we have seen, in times past they have been active in passing laws to protect the environment. Yet conservatives increasingly accede to the amoral rubbish the extreme libertarians spout – even though it means hastening the destruction of the natural world that they grew up in and value so dearly. For politics as a whole, the result is that without the bipartisan consensus of times past, action on conservation is all too often stalled and stymied. Things have become even more vexed with the advent of climate change, for special interests have so muddied the public's understanding of this issue that it has become the focus of a reaction that threatens to carry other environmental concerns out the door with it.

The antagonism to environmental care which the far-right libertarians are infecting their political allies with, and that they are transmitting to society as a whole, goes much further than the loss of a few species or forests. In their efforts to discredit conservation they've thrown out science as a guide to action, and that has dangerous ramifications indeed. In the case of climate change, it has led many on the far right to an outright rejection of the scientific method, even though every academy of science and relevant institution that has issued a statement on the matter supports the scientific consensus. In the modern world, any society that rejects science is in grave danger. Indeed, science has arguably created modern society, and society cannot be maintained without it. Furthermore, it is science that gives us the capacity to recognise environmental damage, including the damage that is ultimately inflicted on our own bodies and those of our children.

Political opportunism has also played a role in the increasing polarisation of attitudes towards the environment. As the two-party system has weakened, both sides of politics have found themselves courting minor parties and special interests. In New South Wales it is the Shooters and Fishers Party that finds its policies being implemented, courtesy of the O'Farrell government, while in Victoria it's the loggers, some duck shooters and the haters of wind power who are being pandered to by the Baillieu government. Federally, of course, it's the climate denialists who are getting their day in the sun under Abbott's Coalition Opposition.

Despite it all, to re-read Marshall's *Great Extermination* is to be reminded of how very far we have come as a nation in understanding and conserving our natural heritage. At the time the book was written, Australia's national parks system was embryonic, as was scientific understanding of the state of our biodiversity. The total number of Australian bird and mammal species had yet to be established, and only the most rudimentary reckoning of the number of species that had become extinct since 1788 was available. Today we know that eleven species of Australian marsupial have become extinct over the last 200 years; but in *The Great Extermination* (published in 1966) only six species were listed as extinct – and of those, populations of two would subsequently be rediscovered.

At the heart of Marshall's magnum opus is an eloquent and heartfelt plea for Australians to create a continent-wide series of national parks and wildlife reserves. At a time when the payment of bounties for the scalps of now extinct marsupials was a vivid, living memory, it was a vital first step, and yielded precious fruit. Today, national parks and nature reserves cover around 13 per cent of the country. While that's not enough, it is a splendid start, and the creation of the national parks system must surely be seen as the principal environmental achievement of the past half-century. Looking back at the forty-six years since the publication of *The Great Extermination*, I see encouraging evidence that Australia's great experiment is progressing, even if not fast enough to prevent terrible losses.

The clearest evidence of our failure is that Australian species continue to slide towards – and into – extinction. It is now clear that while establishing a national parks system was important, in a place like Australia it is not sufficient to preserve our biodiversity. If we are to approach this task seriously, it's time for Australia's governments and people to take the next step, which is to manage these parks effectively. Yet there are indications that the conservative governments now taking power in Australia are moving away from that aim. Instead, they are, to our great disservice, increasingly using the environment as a political football.

Starting in 2010, a series of conservative state governments has been elected around Australia, and it seems likely that by 2014 such governments will rule almost everywhere in the nation. The actions of the conservative governments elected to date give us some idea of what we can expect as the right sweeps to power. The first to take office was Ted Baillieu's Coalition government in Victoria.

Shortly after being elected, in early 2011 the Baillieu government announced that it would permit the grazing of cattle in Victoria's Alpine National Park. The rationale was to conduct a "scientific" trial to determine whether grazing would lower the risk of fire in the alpine regions. On the face of it, the proposition might seem reasonable; but, as Victoria's National Parks Association wrote, the idea was "deeply flawed, often misleading, and short on crucial information." Although dressed up in the garb of science, the initiative had no scientific basis or support, and a year after it was proposed, Tony Burke, the federal Minister for the Environment, put an end to it – on the basis that it was not scientific and was likely to damage alpine environments.

Victoria has also seen new initiatives with regard to duck shooting. Despite the venerable history of the state's duck shooters as environmental custodians, today they are vilified by a small group of animal-rights activists, and their activities are viewed with suspicion by many urbanites. In a dubious gesture to the shooters, the Baillieu government transferred responsibility for regulating the shooting of native waterfowl from the Department of Sustainability and Environment to the Department of Primary Industries. Whatever impact this has on the birds themselves (and it may have little), it clearly signals a shift in attitude. As the responsibility of the Department of Sustainability, waterfowl were viewed as part of Victoria's natural diversity and deserving of wise management. Their transfer to the Department of Primary Industries, an organisation more used to dealing with animals as units of rural production, indicates their new status as a commodity. The duck shooters of Jock Marshall's day would have been appalled.

Other petty changes designed to appease the anti-environmental right are being made in Victoria. In 2012, the state government began allowing the collection of firewood on public land without a permit. Fallen timber is a crucial habitat for many native species, and removing restrictions could affect a multitude of them. Yet the Baillieu government sees this erosion of our natural heritage as preferable to requiring individuals to obtain a simple permit. Again, it is a small but telling shift. Furthermore, it seems likely that under a Coalition government Victoria will not get a Murray River national park. And, of course, the government's new wind energy policy effectively prevents the development of the state's splendid wind resources, so denying it an important new industry.

The Baillieu government's most recent environmental announcements have managed to mix the farcical with the tragic. Like other state governments, it has announced massive retrenchments of its national parks service staff, yet somehow, within weeks of announcing the cuts, it found money to fund a search for the Gippsland black panther – a beast that deserves a place beside the Loch Ness Monster and Abominable Snowman as a product of the fervid human imagination. The fate of the forests that are to be searched for the mythical creature is, however, no joke, for it is in the area of forestry policy that the Baillieu government's impact on biodiversity is likely to be greatest.

East Gippsland is home to an exceptional range of fauna and flora, some of which is found nowhere else. A species of particular significance is the long-footed potoroo (*Potorous longipes*). One of the smallest and rarest members of the kangaroo family, it is restricted to a few dense forests in the region, and is listed as endangered. It's particularly vulnerable to logging, in part because foxes can use logging trails to gain access to its habitat. In 2005 loggers mistakenly felled 400 square metres of likely potoroo habitat in Errinundra National Park, and logging continues in prime habitat in many forestry stands at Brown Mountain.

In 2010 a landmark court case (brought by Environment East Gippsland and discussed above) found that, contrary to its legal obligations, the state agency VicForests had neglected to survey for potoroos and other rare and protected wildlife in forests to be logged. In response, the Baillieu government developed a strategic forest working plan that proposes to extend logging contracts from five to twenty years, and to allow the burning of forests as biomass for electricity generation. It is also attempting to change the *Flora and Fauna Guarantee Act* (which protects endangered species in Victoria) so that the Secretary of the Department of Sustainability and Environment can exempt logging interests from the need to survey for endangered species in logging coupes. Under the proposal, this could occur if the departmental secretary deems that the endangered species in question exists in sufficient numbers elsewhere. This is ridiculous: endangered species are endangered species – the loss of a single individual can be a step towards extinction for any of them.

Just how desperate things have become in Victoria is evident from the resignation in September 2012 of one of Australia's most eminent biologists from the Leadbeater's possum (Victoria's state animal emblem) recovery team. Labelling the situation an "absolute disgrace," David Lindenmayer said that he could no longer participate in a system where measures required under the action plan for the possum were not being implemented, thereby dooming the species to extinction. According to the *Age*:

> Despite the recovery team agreeing to make the protection of remaining habitat a priority, Professor Lindenmayer said the three years since had failed to produce as much as an "action statement."

Lindenmayer added that the key threat to the possum is the Baillieu government's new twenty-year logging contracts:

> Locking in twenty-year contracts for the forests in Victoria will lock in the extinction of Leadbeater's possum and I can't be involved in a recovery team that has no effect on environmentally bankrupt policy.

Many in the environmental movement feel all of these changes acutely, and some of them are indeed deadly serious. The fundamental problem, however, is that our governments *are not being held to account for their responsibility to protect our biodiversity.* And that, of course, means requiring governments to set clear goals, have the means to monitor changes to the environment, report back to the community on progress, and be penalised for failure. This is the real battlefield, and in my view it is where society as a whole needs to demand more of its elected representatives.

Some, of course, will argue that it's too expensive to save endangered species. But recent advances have shown that we can save many more species, at far less cost, than we do at present. The principle of accountability applies here too: recently the University of Queensland's Liana Joseph and her colleagues have demonstrated that setting priorities in a rational and transparent way is crucial. Their approach has already been adopted in New Zealand, where they have been able to secure three times as many endangered species as previously.

In March 2011 the Coalition government of Barry O'Farrell came to power in New South Wales, and its first order of business was putting on an ideological sideshow to appease the extremists. In New South Wales, however, more was politically at stake than in Victoria, because the Coalition relies on the support of the Shooters and Fishers Party to pass legislation through the state's upper house. One of the new government's first acts was to pass legislation, proposed by the Shooters and Fishers Party, to ban the creation of new marine parks and extensions to marine sanctuaries for five years. The madness of this action defies belief. The lesson learnt from fisheries management, both globally and in Australia, is crystal clear: if fishermen hope to catch more and larger fish, they need more marine reserves, not fewer.

With ever more efficient fishing methods, and ever more people to feed, the protection of fisheries stocks and marine biodiversity is a matter of urgency. The data is unequivocal: 90 per cent of the world's fisheries are overharvested and in decline, while 42 per cent of Australia's fisheries are overexploited or of unknown conservation status. Many of our marine ecosystems, such as coral reefs, are severely stressed by pollution and climate change in addition to overfishing, and a carefully regulated fishing industry is key to both their survival and that of the fish. The fact that the Shooters and Fishers Party was allowed to shoot itself – and the people of New South Wales – in the foot in such a spectacularly idiotic fashion makes this one of most shameful examples of pandering to the monstrously ignorant that I can think of.

In 2012 the O'Farrell government again granted the Shooters and Fishers Party a major victory over common sense. It announced that it was giving Robyn Parker, the NSW Minister for the Environment, the authority to permit hunting of feral animals in national parks, seventy-nine of which would be open to hunters. The Coalition needed the support of the Shooters and Fishers Party to pass the legislation required to sell off the state-owned electricity generation assets (a move, incidentally, which the O'Farrell government had trenchantly opposed while in Opposition). Depending on how the hunting of ferals in parks proceeds, it may do no environmental harm. But without rigorous scientific monitoring the shooting is unlikely to benefit native species and may even get in the way of managing the state's national parks – for example, if it leads to the ending of feral animal–trapping programs to leave more game for shooters.

On 17 July 2012, further changes to the management of NSW national parks were announced when Minister Parker told journalists that 350 jobs would be cut from the state's environment office. As Ben Cubby and Josephine Tovey of the *Sydney Morning Herald* reported:

> The public service cuts, which amount to nearly 12 per cent of all workers in the Office of Environment and Heritage, mean some national parks will go unstaffed, and visitors will be forced to utilise a "self service" culture …

Just how this will affect the proposal to allow hunters to shoot feral animals in national parks is unclear. Presumably at least some monitoring of hunters (some of whom are as young as twelve) is required; but if national parks staff are not available to do it, who will?

As in Victoria, these grotesque sideshows have roused ideologues on both sides, obscuring the obvious truth that governments must be held accountable for the protection of the nation's biodiversity. Until the people of New South Wales demand clear goals in regard to the protection of their state's biodiversity and the management of its reserved lands, and hold their government to account, the circus will go on.

Incidentally, Australia's national parks and wildlife reserves are not as secure as many believe. In Western Australia, for example, national parks and A-class nature reserves can be resumed for mining or other purposes if the state cabinet gives approval (for national parks), or if the minister decides a project should go ahead (in A-class nature reserves). Barrow Island provides just one example of this. Barrow is arguably Australia's most important "island ark" for endangered biodiversity, yet in 2009 the construction of a new gas-processing hub began on the island, greatly expanding the industrial zone at the expense of natural habitat. In addition, it is planned to re-inject carbon dioxide drawn from the gas piped from the offshore Gorgon gas field into rocks that underlie the island.

Western Australia is not the only state where animals and plants lose out when mining and biodiversity come into conflict. In South Australia mining can occur in national parks under certain circumstances, and in other states mining often takes priority over nature protection. Even under Labor governments with a strong green bent, national parks are not always safe. In 2010 the Queensland Bligh government began the process of de-gazetting a large part of Mungkan Kaanju National Park on Cape York Peninsula, with a view to giving the land back to its traditional Aboriginal owners. Delays meant that the hand-over actually occurred under the Newman government. The Bimblebox nature reserve near Alpha, Queensland, is another area whose status is under threat. It is home to three threatened bird species, but if coal developments in the Galilee Basin go ahead, half of it will disappear into the massive pit of the Waratah Coal mine, leaving the remainder vulnerable to subsidence and groundwater disruption.

There is no upper house in the Queensland parliament to moderate legislation, and this, along with Campbell Newman's gigantic majority, makes his conservative government the most powerful and unfettered in the nation. One of its first priorities was the dismantling of environmental protection, particularly anything to do with climate change. But much is happening in the non-climate area as well. Queensland was the last state in the country to prohibit large-scale land clearing, and some rural voters still resent the decision. Recently, Queensland's Minister for Agriculture made a speech in parliament implying that those who illegally cleared their land may not have to pay the fines that have been imposed on them. In the same vein the Newman government looks set to repeal the Wild Rivers legislation, replacing it with a far north Queensland land-use strategy whose details are yet to be determined. It's too soon to know the impact of such changes on species protection, but a softer line on either land clearing or river health could have massively damaging effects.

The most recent government to be elected in Australia is a conservative government in the Northern Territory, which came to power in August 2012. One of its first acts was to restructure the environment department. It was split into smaller agencies, and public servants were expressly told not to use the word "biodiversity" in the new agency's name – the Biodiversity Unit being seen as an annoying impediment to "development." Instead, biodiversity is a responsibility of the Department of Land Resource Management.

In Australia, states have traditionally had primary responsibility for conservation, although in recent decades the federal government has played a growing role. The importance of federal action is underlined by three Gillard government initiatives: the massive expansion of marine parks and sanctuaries in territorial waters, the billion-dollar biodiversity fund, and efforts to resolve the Tasmanian forestry dispute. The principal piece of legislation enshrining federal powers is the Environment Protection *and Biodiversity Conservation* (EPBC) *Act*. It is supposed to provide a legal framework for protecting and managing nationally and internationally important flora, fauna, ecological communities and heritage places. In the last thirty-eight years, it and its predecessors have played an important role, providing a legal framework without which the lower Gordon River in Tasmania and the Mary River in southeast Queensland (home to the ancient lungfish) would now be dammed, our tropical rainforests logged, and many other less well-known environmental assets degraded.

Remarkably, debate about the EPBC Act within the two major parties now seems to be centred on limiting its powers rather than strengthening them. Under its "cutting the green tape" initiative, the Gillard government has expressed a determination to streamline the approval process and delegate authority to approve projects (other than those involving World Heritage areas) to the states. There is no doubt that the Act could be improved; yet what is being proposed is unlikely to lead to better outcomes, but rather to weaker environmental protection.

The Abbott-led Opposition has proposed an even neater solution to the Act's deficiencies. It has announced that if elected it will devolve *all* responsibility for approving projects to the states. Both the Labor and Liberal proposals threaten to sweep away more than thirty-five years of incremental improvement in environmental protection. These protections were devised because the national interest often differs from the interest of a particular state, and states were manifestly failing to safeguard the nation's natural heritage. Yet now many states are struggling financially, and one of their few remaining sources of revenue is royalties from mining. If the states are made solely responsible for biodiversity protection, they will, on occasion, face an intractable conflict of interest.

The EPBC Act was designed principally as a safety net for preventing developments, such as mines or agriculture, that threaten endangered species and areas of national significance, and it remains of critical national importance. But today there are other threats to biodiversity which are far more important in causing extinctions. In my view there's an urgent need for federal legislation that reflects this new reality. A federal policy of zero tolerance to species extinction would be a strong foundation upon which to build protective legislation.

Regrettably, the basic tools that government needs to provide adequate environmental protection are slipping from its grasp. Managing biodiversity starts with a clear understanding of where the threats are coming from – by discerning the various forces at work within ecosystems. This is the work of the scientists, and it often involves an enormous amount of painstaking research – research that can take years or even decades to complete, and which requires long-term, continuous and patient funding to bring to completion. Yet, tragically, the very organisations charged with custodianship of our natural heritage have been allowing their research capacity to wither on the vine. The CSIRO's Division of Wildlife Research, which is part of the National Wildlife Collection, was once a lead organisation in the field. Today, due to a failure to replace key staff, it's little more than a shell, and has been relegated to the status of a marginal player as far as research goes.

Just as dismal is the fate of the various state museums. Long bastions of scientific expertise, particularly in the vital work of taxonomy (identifying and classifying species), they, too, were once in the front line of conservation efforts. The Australian Museum in Sydney offers one example of how low things have sunk in the state museum sector. When I joined its staff in 1984, over thirty researchers were employed, divided between more than a dozen specialist departments which covered much of Australia's fauna. Today there are just twelve fully qualified, full-time researchers left in the institution. Gone are the full-time, permanent curators or research scientists in the departments of reptiles, fish, archaeology, palaeontology, arachnids (spiders), minerals and echinoderms. The Australian public need to know how it is that our precious and unique creatures will be preserved if there is nobody left who is able to identify them.

What then needs to be done if Australia's biodiversity is to be preserved? To understand the threats facing our natural heritage we need to take a journey back in time some 45,000 years, from 2012 to around 43,000 BC, for that was when key aspects of the ecological relationship between humans and their environment first took shape.

Of course it is difficult to know precisely what happened so long ago, and in order to investigate such things we must enter the realm of the scientific hypothesis. In 1994 I published one such hypothesis about what might have occurred when humans first arrived in Australia. Back then, even the approximate date of human arrival was uncertain, with guesses ranging from hundreds of thousands to a few tens of thousands of years ago. Given the lack of data, my hypothesis drew on first principles and observations about how ecosystems respond to the arrival of new species.

The argument was developed in my book *The Future Eaters* and began with the proposition that, just like the first European settlers, the ancestors of the Aborigines had arrived (from Asia) without any knowledge of Australia's ecosystems and fauna. And as they were the first humans (indeed the first primates) ever to reach the continent, Australia's creatures had no experience of a predator anything like them. This naivety was put forward as a key factor in the extinction of over sixty species of mega-fauna – large-to-gigantic marsupials, birds and reptiles (including 95 per cent of Australia's mega-herbivores, species weighing more than forty-five kilograms) – which once roamed this land.

The largest of the megafauna were the diprotodons, rhino-sized mar-supials related to wombats. They shared the plains with marsupials that resembled the extinct ground sloths of the Americas, and one wombat relative that may even have been hippo-like. There were also around thirty species of gigantic kangaroo, some of which had shortened faces and weighed a quarter of a tonne, others being wolf-sized carnivores. There were gigantic wombats, too, which dug enormous burrows, and a leopard-like carnivore related to the koala. Then there were the giant lizards, including the Komodo dragon and a species twice its size, gigantic horned turtles and huge flightless birds. All, the hypothesis stated, had been driven to extinction by human hunting over a relatively brief period – perhaps 1000 years or less.

A number of remarkable discoveries concerning Australia's megafauna have been made over the last eighteen years – discoveries that allow both for the testing of these ideas and the supplementing of some with entirely new understandings. Discoveries of fossils, made in caves on the Nullarbor Plain, show that the megafauna consisted not only of gigantic species, but also of a number of medium-sized marsupials that once inhabited the arid inland. Several were similar in size to the wallabies and kangaroos living today, but had strange skeletons which indicate different lifestyles. One wallaby, for example, had a long, delicate snout and great protective protuberances in front of its eyes, giving rise in the popular press to the bizarre idea that it was a "horned" wallaby. These features may have allowed it to reach deep into prickly vegetation to pluck tender morsels without damaging its eyes.

Most unexpected was the discovery of nearly complete skeletons of two species of tree-kangaroo. We know from studies of invertebrate fossils preserved in the same caves that when these tree-kangaroos lived on the Nullarbor, the region had a climate similar to that of today. Yet nothing on the Nullarbor now grows taller than knee-high. Clearly, the vegetation of Australia's arid centre must have been very different during megafaunal times, even though the climate was similar. Perhaps whole forests of arid-adapted trees became extinct along with the rhino-sized diprotodons and marsupial giants.

The Future Eaters hypothesis included an argument about how the extinction of the megafauna may have changed the vegetation. Most of Australia's soils are old and worn out, and the climate, while predominantly dry, is very variable. In such circumstances, almost all of the available nutrients are held in living things. This means that an abundance of life can only be maintained if the nutrients in plants and animals are recycled swiftly and without loss.

To understand the role the megafauna played in this, consider the fate of a stalk of grass or a leaf, both before and after the megafauna vanished. There are really only three possibilities for recycling it so that its nutrients become available to other living things: either it decomposes, is eaten by something or is consumed in a fire. Each of these possibilities has a very different effect on ecosystem health and structure.

Because nutrients are not visible to us when we look at ecosystems, this can be difficult to understand. One way to envisage the situation is by using an analogy with something more familiar. Imagine a nation that has a fixed amount of money, and think about the fate of its inhabitants when the money supply changes in various ways. First, imagine a situation where the money moves from hand to hand rapidly – for example, every dollar available to the populace is spent once a day. Incomes would be relatively high, meaning that more people could be supported at a greater level of affluence than if the money circulated more slowly.

Next, imagine the situation that would develop if, for each dollar that is spent, ten cents were added to the money supply. The economy would then be supercharged. This is roughly how an ecosystem which is stocked with large herbivores works. Great herbivores are walking composting machines which feed plant matter into the capacious microbe-filled fermentation vats that are their digestive systems, and then shit out copious amounts of dung enriched with animal nitrogen and other nutrients. Because they wander, they also carry nutrients from places like waterholes and spread them out over the countryside. In Australia, their manure was doubtless rapidly buried by armies of now-extinct dung beetles, making its nutrients available almost immediately to new, growing plants. And of course the large herbivores watered the country with their urea-rich urine, contributing in no small part to its fertility.

To understand how Australia changed when the great herbivores were hunted to extinction, we need to imagine an economy where, instead of each dollar changing hands once a day, it is exchanged only once a month, with no ten-cent bonus being paid. With earnings less than one-thirtieth of what they were previously, fewer people can be supported, and at a lower level of affluence, than in the earlier economy. That's broadly analogous to the situation that developed in Australia's eco-systems when the great herbivores were removed, and only microbes in the soil — which act much more slowly than those in heated fermentation vats — were left to decompose the plant matter.

The situation in many parts of Australia was complicated by the fact that the land is dry, and the drier or colder a place is, the longer it takes for microbes to turn dead plant matter into compost. To understand what hap-pened in the drier parts of the continent we must imagine an economy where the circulation of dollars is slowed from once a month, say, to once a year. And we must not forget that in a dry land there's always the chance of fire. If fire consumes the dead plant matter instead of microbes, it's the ecological equivalent of paying a ten-cent penalty, instead of receiving a ten-cent bonus, each time a dollar changes hands. Why a ten-cent penalty? Because the burnt plant matter can be blown away, and rain can wash away the exposed soil, leading to a loss of nutrients. Moreover, essential elements such as sulfur, selenium and nitrogen can be volatilised by fire (and so carried away by the atmosphere), even if they reside deep in the soil profile. Such elements are essential to healthy plant and animal growth, and their loss can have a severely limiting effect. The extent of the penalty paid varies, of course, depending on the nature of the fire. Small, cool fires can minimise the penalty. But large, very hot fires maximise it, and by analogy this can mean the difference between losing one cent, or fifty cents, for every dollar that changes hands.

To summarise, consumption of plants by large animals results in healthy ecosystems because it leads to very fast, efficient recycling of nutrients as well as the delivery of a nutrient and water bonus to the soil. Decomposition is a slower and less efficient way of recycling nutri-ents. Recycling through fire is both slow and inefficient. Worst of all is recycling through large, hot fires, which can lead to permanent impoverishment of the soil. Table 1 summarises these various ways of cycling nutrients through the ecosystem.

TABLE 1. A comparison of the speed and efficiency of various ways of recycling plant matter.

	SPEED OF NUTRIENT RECYCLING	COMPETENCY OF NUTRIENT RECYCLING
Consumption by mega-herbivores	Hours to days	With bonus of minerals/nutrients
Microbial decomposition	Weeks to months	Without bonus
Small, cool fire	Months	With small loss
Large, hot fire	Months to years	With large loss

What happened to Australian ecosystems when the large animals became extinct all those thousands of years ago? The vegetation faced one of two fates: either microbial decomposition in the soil or burning by fire. Moist conditions are required for composting, and in a dry land fire inevitably consumed the lion's share of the dead plant matter.

In order to comprehend the full scope of the changes the extinction brought to Australia, we need to know a little more about the environment that existed before the first human footfall marked the continent. The climate was not much different from that prevailing today, but there are indications that the vegetation was very different. We've seen that the Nullarbor was probably forested, and we can reasonably conjecture that central Australia was covered with a dry but fertile mosaic of shrublands, woodlands and grasslands, which supported a wide diversity of medium-sized to large animals. Other evidence suggests that rainforest, which probably lost its leaves in the dry season, covered much of Australia's north and northeast. Today only tiny pockets of this type of vegetation survive, in the form of the bottle-tree scrubs of Queensland and northern New South Wales, and the vine thickets of Arnhem Land and the Kimberley.

Vegetation can have a great impact on climate, and in the right circumstances rainforests can be rainmakers. In the Amazon today, 80 per cent of the rain that falls comes from moisture transpired by plants. Like the trees of Amazonia, northern Australia's vanished rainforests might have increased atmospheric moisture by transpiration, creating clouds that carried rainfall far inland on the offshore winds that blow during the monsoon. And the shady forest understorey, with its fallen leaves, branches and masses of roots, could have acted as a vast reservoir for moisture during the dry season, allowing rivers and creeks to run for longer into the dry than they do today.

In The Future Eaters I suggested that the destruction of the rainforests followed on from the extinction of the large mammals. Decomposition in the soil could not remove the vegetation fast enough, especially in the drier environments or during dry times of year; and so fires, whether lit by those first humans or the first dry thunderstorms of the season, consumed the fire-sensitive vegetation. In northern Australia today, Aboriginal people carefully protect the last remnants of the rainforest by burning around its edges early in the dry season, providing a firebreak against hot, late-season fires. But 45,000 years ago people had yet to learn to do this, so the fires ate deep into the rainforests. And as the rainforests turned to ash, the rainfall dropped off until Lake Eyre failed to fill, even in the good times. Such changes must have stressed the surviving forests even more, and eventually Australia became what it is today: a continent with a desert heart.

There was, however, a class of vegetable opportunists waiting to take advantage of the disruption. They had emerged on the worst of the continent's soils – the sand-sheet and sandstone areas such as the Sydney sandstone – where there were never sufficient nutrients to support even medium-sized herbivores such as grey kangaroos. In such environments large, hot fires had always been prevalent, and this is where the heathland plants, including banksias and other species that require fire to germinate – and eucalypts, which actually promote fire by producing dry bark and volatile oils – evolved.

As the fires spread beyond the heathlands 40,000 years ago, fire-tolerant species inherited almost all of the continent. And so it was that Australia became the land of the gum tree. Adapted to life on the very poorest of soils, the heathland plants were probably helped in their spread by the impoverishment of Australia's soils through fire. Eroded by the first heavy rains, the bared soils and their nutrients were carried into the rivers, whose mouths silted up, providing habitat for mangroves.

The people living in Australia back then surely noticed the catastrophe unfolding before their eyes. But with the mega-herbivores and rainforests already gone, they had only two choices: either to allow the large, hot fires to continue to destroy what was left of the biodiversity, or to fight fire with fire by lighting many smaller, cooler blazes. This, I think, was how firestick farming began. And over thousands of years the first Australians learnt to perfect the use of fire to breathe productivity into the land when required, and so to protect its remaining biodiversity. So successful were they that by the time Europeans arrived some 40,000 years later, the Aborigines had become the greatest practical environmentalists on the planet.

The environmental history I've just outlined has significant implications for relations between contemporary Australians and their land. It begs the question, for example, of whether there's a role for large introduced herbivores to replace the extinct marsupial giants in Australia's ecosystems. But most importantly, it posits that humans have become the keystone species in the Australian environment. When they withdraw their "ecosystem services" such as managing fire and hunting, Australia's ecosystems will not revert to some pre-human ideal, but will spiral towards ecological collapse. But are the ideas I put forward in *The Future Eaters* in fact correct?

It's important to understand that the Future Eaters hypothesis was made to be tested. Developing and testing such hypotheses is the core of the "scientific method," and as the great philosopher of science Karl Popper showed, hypotheses can never be proved by testing, only disproved. Thus, despite popular misconception, science is not a search for the truth. Instead, science progresses by disproving hypotheses, and over time the great majority of hypotheses put up for testing are indeed proved wrong. Occasionally, however, an idea is tested and re-tested, yet cannot be falsified. Then it is elevated to that most respected of scientific phenomena: a theory. The theory of gravity, the theory of quantum mechanics and the theory of evolution are three such hypotheses which have stood the test of time. While they can never be scientifically "proved" (for scientific testing is endless), they are so resistant to being disproved that they're now time-honoured, and for ordinary purposes we treat them as fact.

The scientific method has made much of recent human progress possible. Yet it's such a counter-intuitive way of thinking that non-scientists almost always fail to understand it – as, indeed, do more than a few scientists. If you hear a scientist arguing a case like a courtroom lawyer, you can be sure that they are not good scientists – for they are trying to persuade you that they know the truth, whereas science in action is rather a search for errors which can be eliminated so that scientific inquiry can progress.

Public understanding is not helped by the very different meanings that scientists and the public give to the same words. Pity the poor scientists trying to convince people of anything when they talk of "theories" as if they are the ultimate scientific achievement, rather than "mere theories" as the public disparages them; and where "falsification" is regarded as an honourable rather than a duplicitous undertaking!

The Future Eaters hypothesis is easily falsifiable. For example, if it were to be shown that the megafauna became extinct before people arrived in Australia – or that they had lived on for a long time afterwards – the hypothesis would have to be thrown out. Likewise, if widespread large fires were found to precede the extinctions – or if changes to Australia's vegetation occurred before the extinctions did – it could also be discarded. Table 2 gives just a few examples of how the hypothesis could be disproved. It also gives the only sequence of events (Column 1) that would allow it to remain not disproved.

TABLE 2. Examples of how the Future Eaters hypothesis might be disproved. (Events are arranged in chronological order from top to bottom.)

NOT DISPROVED	DISPROVED	DISPROVED	DISPROVED
Humans arrive	Megafauna extinct	Humans arrive	Humans arrive
Megafauna extinct	Humans arrive	Increase in fire	Increase in fire
Increase in fire	Increase in fire	Megafauna extinct	Vegetation change
Vegetation change	Vegetation change	Vegetation change	Megafauna extinct

Initially, scientists had difficulties testing the Future Eaters hypothesis because dating technologies were not reliable enough or did not extend back far enough in time. But by the late 1990s new techniques were being developed. When you are working at the limits of a technology or dealing with new technologies, there are many ways in which techniques can yield wrong results. One technique, known as luminescence dating, dates the moment when grains of sand were last exposed to sunlight. But because it is quite common for fossil bones to be exposed through erosion and then reburied, the bones of the extinct creatures buried in the sand might be much older than the date provided by this technique. So the technique is most reliable where articulated skeletons, which cannot be reburied without losing their articulation, are present. In 2001 luminescence dating revealed that the megafauna of the southwest of Western

Australia and southeast Queensland had become extinct around 45,000 years ago, give or take a few thousand years.

Meanwhile one archaeological site, at Cuddie Springs in western New South Wales, presented a major challenge to the Future Eaters hypothesis. It is the only site in the country to contain abundant megafaunal remains which purportedly date to around 33,000 years ago. If this were verified, the site would be proof that the Future Eaters hypothesis was wrong. A number of factors had to be checked, however, before the Cuddie results could be accepted. The dates had not been taken from the megafaunal bones themselves, but from the charcoal and sand that entombed them. And the megafaunal remains were not articulated. In 2009 direct dating of teeth using a technique called electron spin resonance (ESR) showed that the megafaunal remains are much older than the sediments that enclose them, being at least 50,000 years old, while some remains are over 400,000 years old. Hence, because the megafaunal bones and human artefacts differ greatly in age, the Cuddie Springs excavations are not able to provide a robust test of the Future Eaters hypothesis.

In March 2012 a stringent test of the hypothesis was published when researchers investigated lake sediments in northeastern Queensland. Lynch's Crater is an ancient volcanic structure that has been slowly filling with sediment over the past 130,000 years, and analysis of pollen, charcoal and other materials provides the most detailed environmental record for the period covering megafaunal extinction in all of Australia. The key breakthrough in the new study concerned a fossilised fungal spore known as *Sporormiella*. It is produced by a kind of fungus that grows only on the dung of large mammals, and studies from Madagascar and eastern North America show that it declines when megafaunal species become extinct, and increases again when cattle are introduced. So we can be confident that the abundance of spores in the sediment at Lynch's Crater gives us a fair idea of the numbers of large mammals living around the lake.

Analysis of the Lynch's Crater sediments shows that *Sporormiella* had been abundant until about 43,000 years ago, after which it abruptly vanished, indicating that the megafauna had become extinct. By studying the abundance of charcoal fragments in the sediment, the researchers were also able to show that shortly after the *Sporormiella* disappeared, the incidence of fire increased dramatically. Later again, the kinds of pollen present began to change. Before the frequency of fire increased, the area had been dominated by a fire-sensitive type of rainforest, but as the amount of charcoal in the sediment increased, the pollen from fire-sensitive species almost vanished, and the pollen of fire-tolerant or fire-promoting species such as eucalypts increased.

As we have seen, just one sequence of events – megafaunal extinction, followed by increased fire, followed by an altered vegetation – would leave the Future Eaters hypothesis standing (Table 2). And that is precisely what the researchers documented. Despite passing this test, two crucial further tests of the hypothesis remain to be done. The first involves detection of a "smoking gun": direct evidence that it was indeed humans who exterminated the megafauna. What sort of breakthrough could yield such proof? Perhaps only the documentation of dozens of skeletons of megafauna with butchering marks on the bones from across the continent could do that. But as yet we have not even a single bone with unequivocal human-made butchering marks on it. In fact, we have only about thirty sites of any kind with articulated megafaunal remains that date back to within the last 100,000 years. So, if the extinctions did indeed all occur within a period of 1000 years or less, the chances of finding even one site where humans and megafauna overlap are less than one in three.

The second test of the hypothesis involves determining whether the sequence of events found at Lynch's Crater applies continent-wide. After all, you can't extrapolate continent-wide changes from a single site. Scientists are currently working on this test. Oceanographers have taken cores from marine sediments right around the Australian coast, and these sediments are known to contain *Sporormiella*. We should know in a few years whether the situation at Lynch's Crater was typical of the continent, or not. If it is indeed typical, there will be little alternative but to accept the

Future Eaters hypothesis as the dominant theory, and to apply its lessons to contemporary land management.

I have dealt with the Future Eaters hypothesis at length because it has profound implications for the management of Australia's ecosystems. But to close this discussion, let me return once more to the lost world of the megafauna.

When I first learned about the vanished giants, I yearned to know what they looked like. Europe, of course, has its splendid cave paintings of woolly mammoth, rhino and giant elk. But they were made tens of thousands of years after Australia's megafauna had become extinct, and there was no evidence of such art in Australia.

No one knows why both the Americas and Australia lack the rich European-style art galleries crammed with depictions of ice-age giants. But in Europe the great beasts and humans co-existed for tens of thousands of years. Did the great giants of the Americas and Australia become extinct so quickly that the stone-age frontiersmen did not get a good chance to depict them? Or perhaps the roaming, frontier culture of the Australian stone-age hunters was not conducive to the development of art − as was the case on the historic American frontier. I had long given up hope of ever seeing an Australian Altamira, Lascaux or Chauvet when, in 2009, news was published of just such a discovery.

There had been many earlier claims of discoveries of Aboriginal drawings that supposedly depict Australian megafauna, but all are dubious. Most of these drawings are of emu-like creatures or large kangaroos that may represent poorly executed attempts to depict living species, while others are of fantastical beings that resemble no creature living or dead. But the artwork published in 2009 was instantly recognisable as a marsupial lion (Thylacoleo carnifex) − the largest and most distinctive warm-blooded predator ever to stalk our land.

The image was discovered in 2008 by Tim Willing, a tourism operator who spotted it in a rock-shelter in the remote Admiralty Gulf in the northern Kimberley. Circumstances were such that Willing was able to take just three pictures before he had to leave. When I first saw one of the photographs, the hairs on the back of my neck stood on end. There, executed in red ochre, was a life-sized depiction of the great cat-like predator – a sexually aroused adult male, in front of which dangles the tail tip of another individual – a female, perhaps – that the male is following. Regrettably, almost all of this second image has been lost, with only the tufted tail-tip remaining. Aboriginal artistic styles have changed over time, and the marsupial lion is executed in the oldest style yet identified.

The marsupial lion is unmistakable. A relative of the koala and wombats, it was a leopard-sized hunter that lacked canine teeth, and instead killed with the aid of a huge, sheathed claw on its thumb. Courtesy of a careful but long-forgotten artist, we now know that it had powerful forequarters, a large box-like head with triangular ears, a long, tufted tail and an evenly striped back. But by far and away its most striking feature is its extraordinarily muscular forearms, one of which is depicted crooked – like a man about to arm-wrestle – ready to strike with its lethal claw.

When I first saw this precious survivor of an unimaginably distant age, I felt as if I was looking into a lost world through the eyes of a person who was there. I can imagine the pair of marsupial lions that the human had watched, winding their way through the rocky gorges of the Kimberley in a long-lost mating ritual. And I am eternally grateful to the artist for taking the time to depict the creature so faithfully, and in such a naturalistic style. But the artwork raises as many questions as it answers. How old is it? Was it drawn in the first thousand years following human arrival on the continent, or could the pouched lion have survived for far longer than Australia's other giants in the rocky wilderness that is the north Kimberley?

At the time I wrote *The Future Eaters* I didn't consider the possibility that any of Australia's megafauna might have survived the initial human onslaught. But the drawing has made me think I might have been wrong. It seems so unlikely that an image made in the first millennium of this continent's settlement could have survived. Furthermore, in other parts of the world large carnivores have survived far longer than the giant herbivores that they preyed upon. In north Africa, for example, the Barbary lion – the top predator of the Atlas mountains – survived into the twentieth century, many hundreds of years after all the mega-herbivores of the region (documented by the Roman natural historian Aelian in the second century AD) had become extinct.

The Barbary lion was not like its Serengeti relatives, for it had given away life in a pride and become a solitary hunter which fed on gazelles and other small game. It became extinct in the wild in 1922 – at a time when modern firearms were becoming available. But its genes survive. The King of Morocco had long kept a captive colony of the beasts, which were occasionally fed with the monarch's enemies, Roman-style. Sometime in the twentieth century a few of these man-eaters were transferred to European zoos and circuses, where their descendants survive to this day. While in captivity they bred with other strains of lion, creating hybrids, and a project now exists to re-create the Barbary lion by selectively breeding from these descendants, with the aim of eventually releasing the beasts into a national park in the High Atlas.

North America provides another striking example of a great predator that outlived its larger prey. Radiocarbon dating has demonstrated that the giant, hyper-carnivorous short-faced bear (*Arctodus simus*) of North America stalked the plains of the southwest until around 10,000 years ago – at least 1000 years after the last mammoth, horse, ground sloth and any other large herbivore became extinct continent-wide. It must have subsisted on the few bison herds that survived the extinctions. Australia's Kimberley region abounds in wildlife, including rock wallabies, euros and other smaller vertebrates. Perhaps it's not outlandish to imagine that a few marsupial lions clung to survival in the rocky fastness there, for thousands of years after the last diprotodon and short-faced kangaroo were swept into extinction.

Rock artwork depicting a marsupial lion, discovered in 2008
in Admiralty Gulf (photo: Tim Willing).

The arrival of the First Fleet in 1788 initiated a period of rapid change for Australia's environments. Brown rats (*Rattus norvegicus*) probably disembarked at Port Jackson along with the first convicts, and cattle had escaped into the bush by 1789. These first feral species spread quickly, and within a century they'd been joined by goats, foxes, pigs, cats, black rats, rabbits, hares, horses, several species of deer and a dozen or so bird species. Today the tide of introduced species continues to swell, making Australia home to a vast array of ferals – from starfish to carp and camels to cane toads – and each is having its own particular impact.

Paradoxically, some biologists argue that Australia's environmental problems are so bad that they can only be solved by introducing yet more exotics. One of the most controversial of such calls was made by Professor David Bowman from the University of Tasmania, who in February 2012 published an article in *Nature* (the world's leading science journal) suggesting that if Australia's ecosystems are to be returned to health, then elephants should be introduced to the outback.

His argument was that elephants would eat enough vegetation to suppress the vast wildfires that currently inflict so much damage on Australia's inland ecosystems. The discussion that followed included many acerbic objections, one coming from an African wildlife expert who pointed out that elephants are difficult to manage. Whether Australian cow-cockies would be up to the task was, to his mind, questionable. But the most fundamental objection concerned the mismatch between elephants (and indeed any introduced species) and the Australian environment.

As explained above, Australia's megafauna co-evolved with the continent's vegetation for millions of years. The strange "horned" wallaby from the Nullarbor demonstrates that some plants defended themselves from being eaten by megafauna with thorns or a thicket of hard twigs, and some marsupials had adaptations that allowed them partially to circumvent these defences. Other plants doubtless used chemical defences to repel herbivores, and over the ages a balance was struck. One of the great problems with introduced herbivores is that they may not be deterred by such defences, and so can chew through Australia's vegetation as if it were a bowl of salad. Moreover, in Australia such herbivores lack predators, which means they can proliferate to the point where they degrade the ecosystem that feeds them.

I am not saying that some carefully controlled introduced herbivores may not benefit Australia's inland ecosystems by going some way to restoring the "dollar spent a day, with a ten-cent bonus" type of ecosystem that existed here in times past. It's just that we're a long way from understanding what species they might be, and how they might be managed to achieve the desired outcome. Moreover, it's self-evident that ill-conceived and poorly monitored programs, such as the proposed trials of cattle grazing in the Alpine National Park, are no substitute for such studies.

There are a few species of megafauna which, while extinct in Australia, have survived elsewhere. The Komodo dragon exists today only on the island of Flores and nearby smaller islands, yet fossils show that it was once widespread in Australia. Why did it survive on Flores? For around a million years the island was home to a pygmy human-like creature known now as the Hobbit (*Homo floresiensis*). Perhaps hunting by these metre-tall creatures (about the height of a three-year-old human) relieved the Komodo dragons of some of their naivety regarding two-legged human-like predators, and so they were wary enough to survive when the first humans arrived around 12,000 years ago.

A second such species is the long-beaked echidna of New Guinea. Fossils of similar creatures have been found throughout Australia in deposits 45,000 years or more old. The New Guinean long-beaks are gigantic (up to a metre long) relatives of Australia's common echidna, but eat worms and possibly beetle larvae rather than termites. Presumably they survived in New Guinea because the dense forest and difficult terrain made them more difficult to hunt there. The western long-beaked echidna (*Zaglossus bruijni*) inhabits the New Guinean lowlands, and may be able to survive in seasonally dry environments. It is severely endangered by human hunting in New Guinea, but might thrive if introduced into northern Australia.

Because both the Komodo dragon and something like the long-beaked echidna were part of Australia's environment for millions of years, we can be more confident that their reintroduction would not disrupt ecosystems, and would indeed fill a vacant niche. The same is true for the Tasmanian devil, which only vanished from the mainland of Australia around 3000 years ago.

We now need to turn to a discussion of direct human impacts on the environment. Since the first days of settlement Europeans have been active, felling forests, planting crops, hunting and altering watercourses. And doubtless these activities have had a great impact on Australia's ecosystems. Australians often imagine that these are the most important changes that have occurred to the environment, and that they alone are responsible for the extinction of native species. But in fact what the Aboriginal people were prevented from doing by the Europeans was equally important.

Forty thousand years is a very long time. Longer than humans have been present in western Europe; three times longer than the entire human history of the Americas. Long enough, indeed, for a human culture to become the keystone in the environment of a continent. The Aborigines acted as a keystone in Australia by carefully burning the vegetation that was once eaten by the megafauna, and by regulating the abundance of the remaining species through hunting. Take a keystone out of an arch, and the structure collapses. And so, when the European settlers began to disrupt Aboriginal land management, they removed the human keystone that lay at the heart of Australia's ecosystems. Environmental collapses can occur on the timescale of decades or centuries, and the consequences of this particular keystone removal are still being played out today.

A breakthrough in our understanding of the importance of Aboriginal land management to the survival of many Australian species was made when researchers examined what has happened in the Great Sandy Desert over the past half-century. This vast region was home to the Pintupi people, and it has a unique history that makes it a particularly valuable natural laboratory. Not only was it the last place in Australia where Aboriginal people pursued traditional lifestyles unaffected by Europeans, but it was photographed from the air in great detail as early as 1953.

The aerial photographs were taken by the military because the British and Australian governments wanted to use the region for testing rockets in a program known as the Blue Streak rocket project. They reveal a landscape patterned into a fine mosaic of vegetation, in various stages of recovery from fire. This had resulted from the burning of small patches – most only a few hectares in extent – by the Pintupi. In the 1980s, some of the people who had done the burning were interviewed by scientists who wanted to know how they burnt, and why. The Pintupi said that they burnt every day, and that while fire served a variety of purposes, hunting was the main one.

The last of the Pintupi to follow their traditional lifestyle walked out of the Great Sandy Desert in 1985, so ending 40,000 years of indigenous fire management. But the system had begun to break down earlier than that. Aerial photographs taken in 1973 reveal that there were then too few people to maintain the tight mosaic of small burns, so larger fires were beginning to break out. By 1981, large, hot fires had gained such a hold on the landscape that four huge ones burnt 90 per cent of the study area. What had happened? As the Pintupi left the desert, the vegetation built up until large fires became inevitable. For the mammals and many other desert creatures, the results were catastrophic.

To understand how catastrophic, imagine being a creature the size of a bandicoot or a small wallaby in a harsh desert environment. Each day you must find shelter from the climate and predators, and each night enough food to survive. Where fires are small, you can easily move from an unburnt patch that provides shelter to a freshly burnt one that provides new green shoots. But if a great, hot fire burns through your habitat, even if you survive the flames by sheltering in a burrow, you're likely to emerge to a devastated field of ashes that stretches to the horizon all around. You'll either be picked off by a predator or starve. Some of the wildfires that erupted in central Australia following the end of Aboriginal fire management were large enough to burn through three Australian states, and to be seen from space. And with them went the fertility of the soil, the variety of vegetation so many creatures needed to survive, and the tight mosaic of old and newly burnt vegetation that provided food and shelter. Now the fires have become so large and hot that even some of the fire-sensitive desert vegetation, like mulgas and some desert eucalypts, are endangered by them.

For a while there was hope that some of the mammals of the Pintupi homeland could be saved. Andrew Burbidge is a mammalogist who told me that when he met some Pintupi people who had recently come in from the desert, he was astonished to learn that they had regularly hunted and eaten a number of mammal species that scientists assumed were long extinct. Andrew excitedly arranged an expedition to take the people back to their tribal homelands, hopefully to find some of the long-lost mammals. The Aborigines, too, were excited at the prospect of returning home, as they had not seen their country for a decade or more. But almost immediately upon arriving, they became dismayed. "All gone," they said. There were no tracks of any of the creatures in the sand, and the vast field of uniformly aged vegetation that had developed disgusted them. It needed to be "tidied up," they said. As soon as they got out of the vehicle, they began lighting fires.

As I explained in The Future Eaters, I think that there is a direct link between the extinction of Australia's megafauna and the historic extinctions of creatures like the bandicoots and wallabies that recently vanished from the Pintupi country. After the great herbivores vanished, wildfire should have led to the extinction of the smaller mammals. It was only Aboriginal fire management which prevented that. The management was an act of environmental stewardship that involved the lifetime efforts of countless millions of long-forgotten Australians, who passed down, in an unbroken string of learning and action, daily practices that became the keystone in Australia's drier ecosystems. Every species given life by that stewardship should be looked upon as the most precious gift imaginable – a gift of rare beauty, diversity and ecological health bequeathed by Australia's indigenous people to those alive today.

Tragically, the kinds of extinctions seen in the Great Sandy Desert in the 1970s are now devastating vast regions of the continent, including some of Australia's most valued national parks. The situation is particularly dire in northern Australia, where, over the past decade or two, the second extinction wave is beginning to break and is sweeping away the small to medium-sized mammals as effectively as the first wave did in southern Australia a century ago.

At 19,804 square kilometres, Kakadu National Park is arguably the jewel in the crown of Australia's reserve system. World Heritage-listed, it received $18 million for operating costs in 2008–09, much of which goes to managing the influx of tourists. Yet, unless it is to become another marsupial ghost town, more needs to be spent on biodiversity protection. Between 1995 and 2008 the abundance of small mammals found in the park declined by 75 per cent, and a third of the species that were recorded there in 1995 can no longer be found, and appear to be locally extinct. One, the brush-tailed rabbit-rat (the last surviving relative of the white-footed rabbit rat of southeastern Australia), may even be extinct nation-wide. As the researchers who reported these dismal findings surmised:

> The current rapid decline of mammals in Kakadu National Park and northern Australia suggests that the fate of biodiversity globally might be even bleaker than evident in recent reviews, and that the establishment of conservation reserves alone is insufficient to maintain biodiversity. This latter conclusion is not new; but the results reported here further stress the need to manage reserves far more intensively, purposefully, and effectively, and to audit regularly their biodiversity conservation performance.

FIGURE 1. Trapping success rates for individual mammals and for mammal species, between 1996 and 2008, in Kakadu National Park.

A. INDIVIDUALS

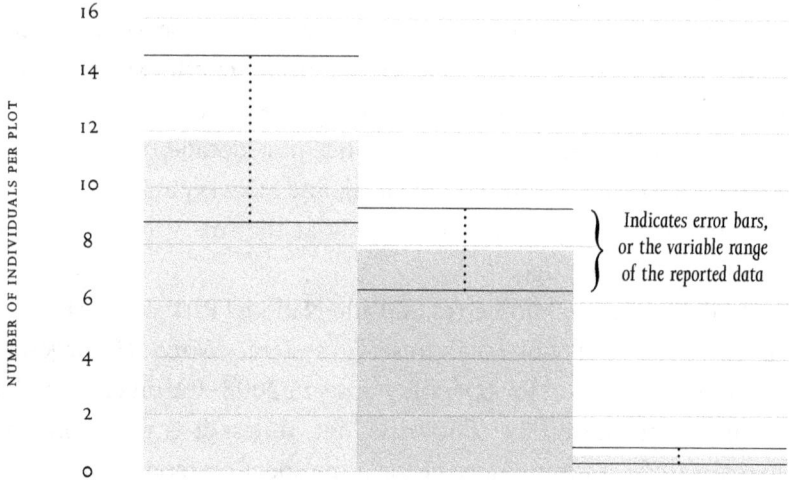

} Indicates error bars, or the variable range of the reported data

B. SPECIES

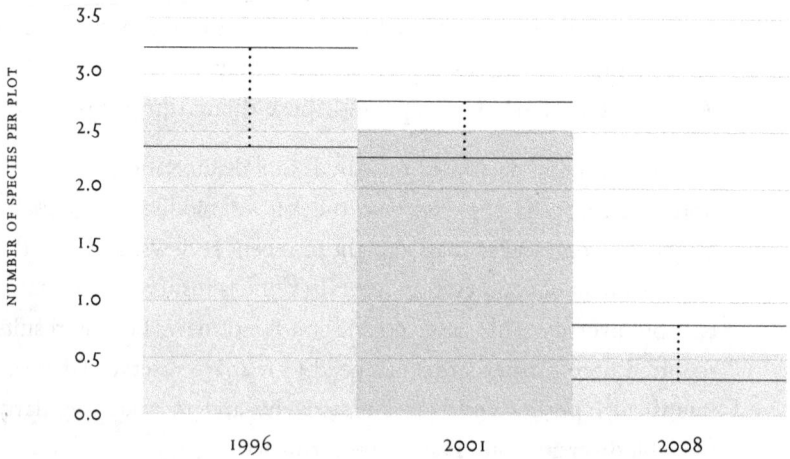

The ongoing extinction of northern Australia's medium-sized mammals, such as bandicoots, quolls, rabbit rats and tree-rats, while catastrophic in itself, is just one symptom of an ecosystem in extinction freefall. As the extinctions continue, vegetation communities are changing and simplifying, while populations of reptiles and birds, such as the Gouldian finch, are also being affected. These shifts are occurring on such an awesome scale and proceeding so fast that they resemble a local version of the dinosaurs' extinction. What has changed in recent decades across the north, including in Kakadu, the most highly protected region of Australia?

The main driver appears to be changes in fire regime, compounded by the presence of feral cats, and shifts in the abundance, in some habitats, of large feral herbivores such as cattle and water buffalo. The water buffalo had been introduced from Asia in the nineteenth century, and by the 1970s were wreaking havoc with aquatic ecosystems. They would crowd the billabongs in the dry season, destroying water lilies and reeds, and stirring up mud until the waterholes became mud wallows, the buffalo using them looking like so many maggots in a sore when viewed from the air.

There was thus good reason to cull the creatures, but the cull may have had unfortunate consequences. While the dynamics of the situation are still being worked out, I gained a small insight into one aspect of it in the mid-1990s. I was filming an ABC documentary when a parks ranger explained to me that a program to eliminate water buffalo from the park had changed the fire regime. Adult water buffalo each eat around twenty kilograms of dry grass per day, such that prior to the cull they removed much flammable material from the Kakadu floodplain. With the buffalo gone, hot fires began to consume the uncropped grass. Fuelled from the floodplain, extremely hot and large fires then began to sweep deep into the surrounding escarpment country.

I recall standing beside one victim of the flames, a gigantic *Allosyncarpia* tree, its eighty-centimetre-thick trunk still smouldering, its once dense canopy a mess of browned leaves and ashes. This giant must have been hundreds of years old, and had been growing in a moist fold in the sandstone escarpment, where it had been safe from fire all its life – until, in the absence of both grazing by water buffalo and Aboriginal fire management, its trunk was burnt through and it collapsed. Tragically, it was not alone. The whole area was being transformed by the enormous, extremely hot fires.

If we are to fully understand what's happening to Australia's biodiversity, we must also consider the assault of cane toads, feral cats, pigs, cattle, horses, donkeys, noxious weeds, and in southern Australia foxes, cats, rabbits, camels, deer and goats. In all, seventy-two vertebrate species have established feral populations in Australia, and when combined with the fires, their varied impacts are making our national parks unsafe for native species continent-wide. Those living close to the land have long understood this, as was recently highlighted in Adelaide at a public meeting convened by Bob Debus to discuss the federal government's billion-dollar biodiversity fund. Debus had been talking about the need to connect up national parks by creating corridors which would allow species to migrate as Australia's climate changed. An Aboriginal elder from Cape York responded that it sounded like a great idea – in theory. But the fact was that many of the national parks in his region were infested with feral pigs, which would use any corridors that were created to spread and thereby inflict even more damage on the environment.

Medium-sized native mammals are critically important to Australian environments. They include the largest burrowers in many habitats, and their burrows provide refuge for many other species. Moreover, the spoil heaps created by digging bring fresh nutrients to the surface, providing important habitat for many ecologically important plants. Rabbits burrow, too, but they also destroy the plants. Moreover, some marsupials are fungus-eaters, and they spread fungal spores that are important to forest health. Others disperse seeds, eat insects that can kill trees if their numbers build up, and distribute nutrients such as phosphorus across the countryside. Take out these vital functions and you end up with sick ecosystems.

What is to be done? At the highest level, it's clear that Australians today must take up the role, forged over 40,000 years, of acting as a keystone species in Australia's varied environments by managing fire, regulating the numbers of feral animals and eliminating weeds. If this is not done, then northern Australia will lose many of its important species, while in the south the last remnants of the medium-sized mammal fauna will be lost. Just where this would lead over time is unclear, but both ecosystem stability and productivity are likely to be affected.

With Australia having over 22 million people in a continent of nearly 8 million square kilometres, it is fair to ask if these things can be achieved within an affordable budget. A partial answer to this question comes from a recent assessment of endangered species in the Kimberley region. A study of 637 vertebrate species showed that 45 mammals, birds and reptiles were likely to become extinct in the next twenty years without action being taken to manage fire, feral animals and weeds, and grazing. Yet the cost of all actions to avoid the extinctions was just $40 million a year – a startlingly small sum given the many benefits that flow from fire management and control of pests in one of Australia's most valuable tourist regions.

We need not rely on theoretical studies alone as a guide to the cost-effectiveness of protecting endangered species in the Kimberley. The Australian Wildlife Conservancy (AWC) is a not-for-profit organisation funded principally through donations from the public. It was established by Martin Copley, whose vision to protect Australia's endangered species has had an enormous impact. (Here I must declare a personal interest: I've been involved with the AWC since its inception, and now serve on its board of directors.)

In little over a decade the AWC has grown to the point where it manages over 3 million hectares. Its reserves are scattered throughout the nation, with particularly significant holdings in the tropical north and centre. On this land, the organisation conserves around two-thirds of Australia's threatened mammal species, and 70 per cent of its threatened mainland bird species. And it manages to do this on an annual budget of around \$12 million – just two-thirds the budget of Kakadu, which occupies an area half the size of the AWC's holdings.

The operations of the AWC vary in accordance with the needs of the region, but the challenges it faces are particularly diverse and urgent in northern Australia. Mornington Sanctuary in the central Kimberley is the headquarters of the organisation's research program, and the place in northern Australia where it has been operating longest. When the AWC commenced scientific management of the property in 2004, the wildlife responded almost immediately. In fact, the abundance of native mammals more than doubled in the plots sampled between 2004 and 2007.

In the early days at Mornington the most immediate task was reducing the impact of cattle and horses by removing them from vulnerable habitats.

FIGURE 2. The recovery of small mammals on Mornington Station in the southern Kimberley, during 2004–07.

(YSD = years since destocking; a quadrat is a square area sampled for mammals; blacksoil, riparian, coolibah and sandseep are habitat types.)

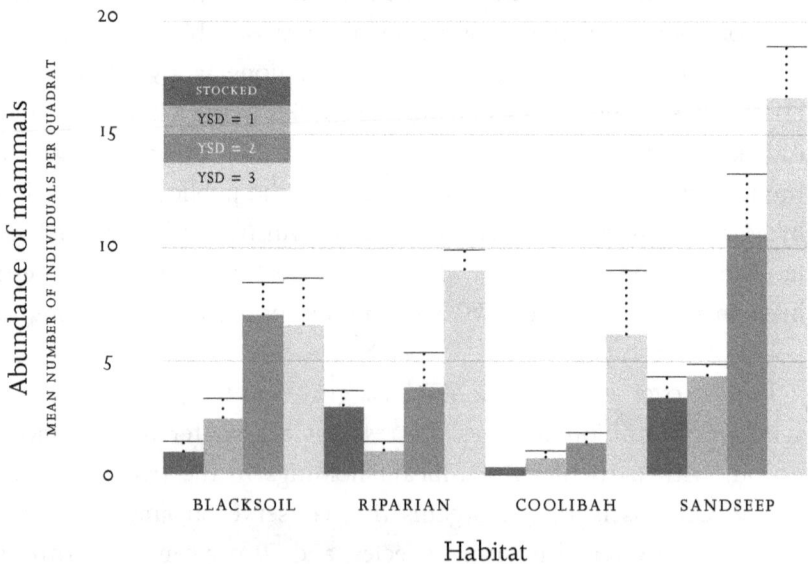

But it was clear that a fire strategy was also urgently required. So, in 2007, the EcoFire project was initiated. Fire is best managed regionally, so the AWC began working with neighbouring properties. As a result, the AWC now manages fire over thirteen large properties with a combined area of 4 million hectares – around a third of the Kimberley. Three other organisations also manage fire in the region: the WA Department of Environment and Conservation, the Kimberley Land Council and the Fire and Emergency Services Authority. It will be interesting to compare their track records when the data becomes available, but it's already clear that the EcoFire program is essential to the continued survival of small mammals, as well as several bird species, in part because it has resulted in a significant increase in the extent of old-growth vegetation (defined as vegetation which has grown for more than three years since the last controlled burn). Moreover, this old vegetation is scattered in a fine-grained pattern across the region.

Partnerships with indigenous people play a vital role in the EcoFire project, as well as in other AWC initiatives. A major step forward in working with indigenous communities was taken when a new agreement was signed with the Yulmbu community in the central Kimberley. As reported by the AWC in 2012, the Yulmbu are:

> subleasing their land (Tableland Station) to AWC for more than forty-five years. It will be managed in accordance with an agreed strategy which encompasses specific targets for feral animal control and fire management. The benefits to Yulmbu are significant: they receive annual income (the sublease payments), training and employment in the delivery of land management programs, infrastructure improvements and a modest, sustainable cattle operation ...
>
> Structured as a private sector project, with measurable performance targets, this innovative partnership will improve ecological health across 3000 square kilometres of the Kimberley and deliver jobs, education, income and other benefits for the Yulmbu community.

FIGURE 3. The estimated number of numbats (*Myrmecobius fasciatus*) on properties managed by the AWC at Scotia (solid line) compared with those in Dryandra Woodland, Western Australia.

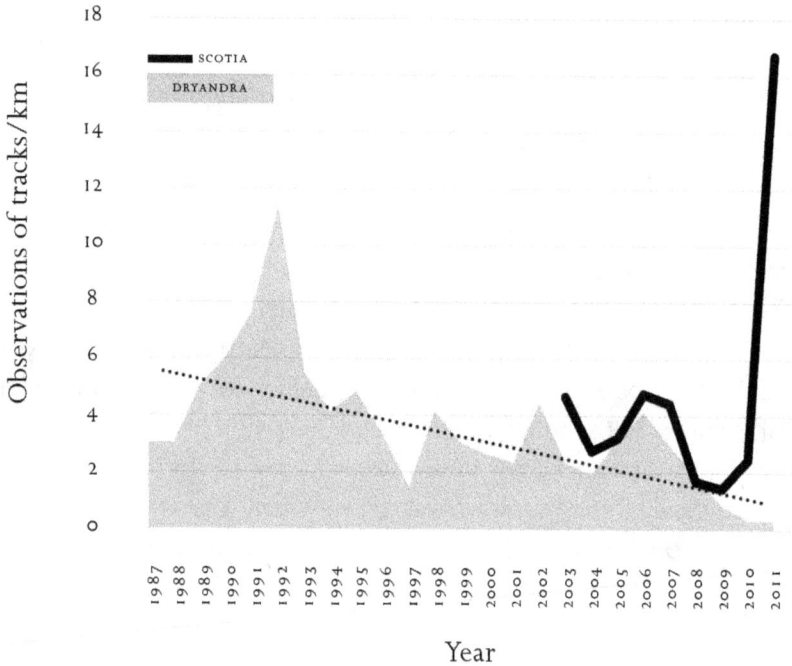

Such a joint management approach, with agreed targets and strategies, has much potential for managing biodiversity nationwide, including in national parks.

In southern Australia, the challenge of environmental conservation is significantly different, and requires other measures. Many of the medium-sized mammals, such as woylies and rabbit rats, that once abounded in the south are now extinct or survive only as remnant populations on islands, where they have lost all wariness of mammalian predators. In these circumstances feral-proof enclosures offer the best method of conservation. This involves fencing areas, eliminating cats, foxes, rabbits and other ferals, and reintroducing the endangered species. These fenced areas

FIGURE 4. The estimated number of bridled nailtail wallabies (*Onychogalea fraenata*) surviving on properties managed by the AWC (principally Scotia) and all other populations from 2001 to 2011.

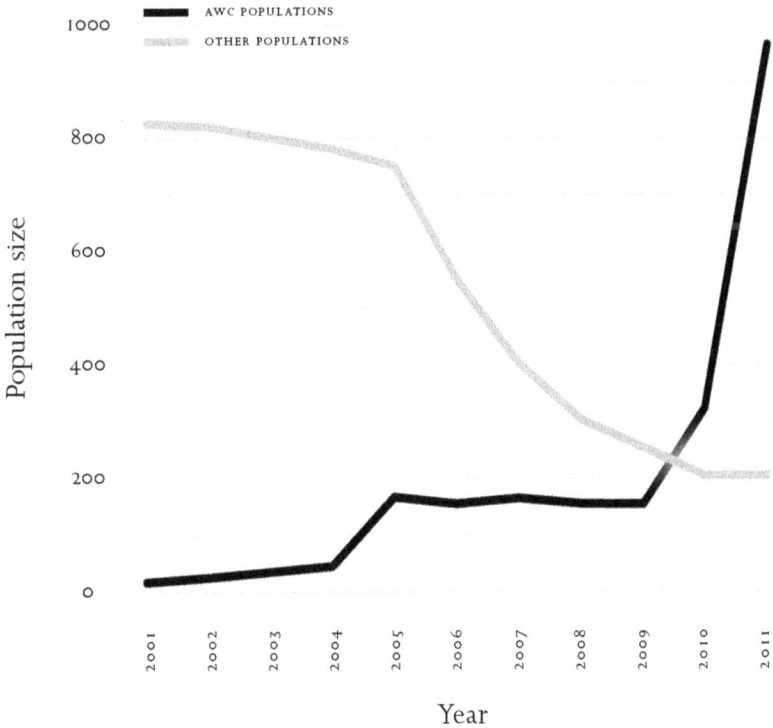

effectively act as arks, keeping the survivors safe from extinction and genetically diverse until better options are developed.

To maintain viable populations of some species, truly enormous feral-proof areas need to be created. At Scotia Sanctuary in western New South Wales the AWC has created an 8000-hectare fenced area – the largest in the country, and vital to species such as the numbat and bridled nailtail wallaby. Fencing involves a significant up-front investment, but over the long term costs compare favourably with alternatives such as ongoing baiting for cats and foxes. Figures 3 and 4 illustrate the fate of both endangered species, which are among Australia's most endangered mammals, on AWC-managed lands (principally Scotia) and elsewhere. Simple extrapolation indicates that unless something changes, the species will eventually become extinct outside AWC-managed areas.

Why has the AWC been so successful at conserving biodiversity? Clarity of vision is important. The organisation's sole objective is the protection of Australia's biodiversity. Moreover, it is committed to creating and maintaining a strong scientific base, which means that it can establish appropriate goals, monitor progress and report accurately to stakeholders. This gives the AWC a very different internal structure from its peers. For example, the AWC devotes around 84 per cent of its total budget to field-based staff and operations, including its science programs. Among other not-for-profit organisations in the environmental sphere, and indeed in comparable government agencies such as parks services, 40 per cent or less of total funding is typically spent on field-based staff and programs, and very little on science. The focus on science is also reflected in staff expertise: the AWC employs twenty-five ecologists (a dozen of these with PhDs) and has about fifteen students, from doctoral to honours level, at work at any time, as well as running a very popular internship program for recently graduated ecologists who are seeking field experience.

A good example of how science guides the AWC is its approach to protecting one of Australia's most spectacular birds, the Gouldian finch. At the time the AWC's Gouldian finch program was instigated, there were thought to be fewer than 2500 adults in the wild, and the species was declining towards extinction. Research showed that this was due largely to changed fire patterns, which had altered the availability of the grass seeds upon which the species feeds, which also meant that during the breeding season birds were becoming stressed by the great distances they had to travel between their nesting hollows and the nearest seeding grasses.

This may sound like a simple finding, but establishing the facts and eliminating other possibilities required a sophisticated program of scientific research that took several years to complete. With research completed, AWC staff started to burn parts of their properties in ways that encouraged the seeding of grasses near nesting hollows of the finches. Today, as a result of the program, Gouldian finches are thriving on Mornington, and blood analysis shows that nesting females are far less stressed than in years past and therefore more likely to survive into a second breeding year. The program also benefits other species that share the finches' habitat, and indeed is emblematic of how protecting a species can lead to healthier ecosystems.

A strong scientific basis means that the AWC can account for its successes and failures. It also means that it is today the only conservation organisation in Australia able to provide reliable estimates of the population sizes of each endangered species in its care. The ability to demonstrate the success of its operations is also a great magnet for staff: talented young people are keen to work for such a body because they can be sure that their work is effective.

The AWC is not the only organisation succeeding in protecting threatened species. In Western Australia, Wildlife Research and Management has had success in protecting the burrowing bettong at Heirisson Prong on Shark Bay, demonstrating that community-based initiatives with some government funding can be effective. Nonetheless, the sheer scale of the AWC programs, and their high rate of success, is impressive.

It is reasonable to ask if not-for-profit organisations can be charged with protecting Australia's biodiversity in the long term. After all, being dependent on donations, they may be thought to have a tenuous existence. But overseas some not-for-profit organisations have been operating successfully for a century or more. Such organisations suffer ups and downs as the economy expands and contracts, but so too do government departments. Indeed, one AWC staffer recently told me that employees there feel far more secure in their jobs than do their government-employed peers.

It may also be argued that the AWC has a particular vulnerability in that much of the land it manages consists of pastoral leases, and so it does not enjoy the legislative protection of national parks and nature reserves. Yet, as we have seen, even lands protected under legislation are not exempt from being resumed for development. And the AWC is increasingly managing land on behalf of others, making it less vulnerable to changes in tenure, and ever more capable of using its expertise to preserve biodiversity in a wide variety of circumstances.

With biodiversity in Australia's national parks in decline, there's a crying need for state governments to reach out to organisations like the AWC for help in managing the problems that they can no longer adequately address acting alone. Were an organisation like the AWC to be given a role in managing feral animals, weeds and biodiversity in national parks, the state could hold it responsible for achieving a clear set of mutually agreed goals. If they were not met, the state could sack the organisation and employ another to do the job.

Some may object that letting private enterprise into the state-owned realm will only lead to a further hollowing out of government expertise. But public–private partnerships are hardly new, and they've proved useful where governments alone lack the means of achieving results. Moreover, the risk to the environment of this initiative failing could be minimised if it were trialled first in national parks or nature reserves where biodiversity values are already low. The truth is that things are now so dire that we cannot afford to persist with business as usual: a change of direction is essential if we're to head off the great impending wave of extinctions. Australia needs several organisations like the AWC, which would compete for funding and the privilege of conserving our endangered fauna and flora.

AUSTRALIA IN THE REGION

One trend that Sir Keith Hancock could not possibly have envisaged when he wrote of Australia's great experiment is the speed at which globalisation is occurring. Indeed, such is its influence today that for many purposes it hardly seems sensible to speak of "Australia" as Hancock thought of it. Instead, Australia often acts as a small, highly distinctive, wealthy and educated element in a globalised economy that is struggling to deal with global environmental threats and challenges. And as the world acknowledges the need for environmental protection, Australia's actions are being closely watched. One failing increasingly difficult to hide is hypocrisy. The pious urgings of so many Australians to citizens of countries far poorer than our own to preserve their rainforests and other instances of biodiversity, while we do so little at home, is increasingly resented. Imagine being a Bangladeshi villager who is encouraged to share their environment with man-eating tigers – by a people so uncaring about their own environment that they can't lift a finger even to protect a small bat or endangered rodent. Our inaction on such matters is swiftly destroying Australia's reputation as a leader in environmental protection.

This loss of credibility is occurring at a time when there's a great need for Australian leadership in conservation, and nowhere is the need more urgent than in our own backyard. Papua New Guinea is our nearest neighbour, and it's fortunate to have retained its great rainforests almost intact into the modern age. With the exception of six medium-sized to large marsupials (wombat-like creatures and giant wallabies), which became extinct when humans first arrived around 45,000 years ago, and the thylacine and two other wallaby species, which vanished around 3000 years ago, the island of New Guinea has been spared species extinction. Outside parts of Africa, and possibly Southeast Asia, it's thus unique in having a largely intact fauna and flora. But over the past sixty years the human population of the island has quadrupled in some areas, and this is putting enormous pressure on its biodiversity.

The threats are most evident in Papua New Guinea's North Coast Ranges, which also provide a first-rate example of the benefits of Australian leadership in nature conservation. At a little over 1500 metres high, the North Coast Ranges are low by Melanesian standards, yet because they were once an island archipelago they are home to a unique fauna and flora. Overhunting by the much-enlarged human population is endangering many of the region's larger species, in particular its tree-kangaroos. And this is a great pity, because the North Coast Ranges are home to three species of tree-climbing kangaroo – more than anywhere else on earth. Moreover, two are found nowhere else, and were only discovered and named in the 1980s when I led a faunal survey of the region. The tenkile (*Dendrolagus scottae*) is a very large black tree-kangaroo, the males of which have a distinctive and not unpleasant odour reminiscent of human male sweat and pine needles. When I named the species in 1985, it was down to less than 100 individuals, which clung to existence on the highest and most rugged slopes of Mount Somoro, south of Aitape. The second species is known to the local people as the weimang (*Dendrolagus pulcherrimus*). It had become extinct on Mount Somoro in the 1940s, but a tiny population persisted at the far eastern end of the North Coast Ranges. With its white face, blue eyes, golden shoulders and caramel and white-banded tail, the weimang is the most beautiful of all tree-kangaroos. It and its close relatives also have the largest brain relative to their body size of any marsupial – making them the Rhodes Scholars of the kangaroo world.

The tenkile, and possibly the weimang, may have been extinct by now if not for the dedication of two young Australians. Just over a decade ago Jim and Jean Thomas, zookeepers at Melbourne zoo, were so concerned about the possible extinction of the tenkile that they established the Tenkile Conservation Alliance (TCA). Then, in 2003, they left their home and jobs in Melbourne to live in the small regional centre of Lumi, on the southern slopes of the North Coast Ranges.

A decade on, the TCA is the most successful conservation organisation in Melanesia. Fifty villages whose lands cover most of the remaining habitat suitable for the tenkile and weimang have agreed to join a hunting moratorium, thereby creating what is effectively a huge conservation area spanning most of the higher parts of the North Coast Ranges. Although tree-kangaroos breed slowly, the tenkile at least is already benefiting from the moratorium. Surveys indicate that its numbers have increased threefold – to an estimated 300 – and tenkiles are now recolonising habitat they've been absent from in living memory. The region inhabited by the weimang is remote, and its response to protection seems to have been slower. But the Thomases have managed to take a pair into captivity, and they have produced two young. Today there is only one old man left alive around Lumi who remembers the weimang roaming the forests of Mount Somoro. Jim and Jean dream of returning the species to the wild in the area, and in this they have the whole-hearted support of all the villagers in the alliance.

It's not just tree-kangaroos that are benefiting from the hunting moratorium. Large birds, from dwarf cassowaries to crowned pigeons and palm cockatoos – all of which were once either locally extinct or extremely rare – are now commonly seen in the high mountains. And despite having foregone the protein that hunting once provided, the people of the region are flourishing too. Caged rabbits (which are a domestic strain and would not survive in the wild in New Guinea) and fish from fishponds – both initiatives funded by the TCA – have replaced the protein once provided by wild animals.

And for the first time since the human population explosion led to widespread deforestation and the pollution of streams, people are again enjoying access to clean water. Between 2010 and 2012 the TCA delivered 243 water-tank sets to 39 villages, bringing reliable fresh water to around 10,000 people. Preliminary reports suggest that this has had a significant impact on water-borne diseases. The TCA is also assisting in the education of children through a scholarships program which pays for school fees, and is creating employment through a ranger program that monitors endangered species and provides progress reports to the villagers.

It's important to understand how unusual such achievements are in Melanesia. There is no equivalent program anywhere in the region, and no other organisation I know of in a developing country has had anything like this degree of success. Again, it is a prime example of saving an eco-system by concentrating on saving a species.

How is it that one Australian couple has almost single-handedly trans-formed the fortunes of a people and the biodiversity of a mountain range while trying to save an endangered species of tree-kangaroo? The answer is simple: the Thomases set clear goals, used scientific methods to moni-tor their progress, and reported back to the people – all the while deliver-ing tangible benefits to the community in exchange for their signing on to the hunting moratorium. They also understood that the people of the region had to be sure that they would be with them for the long term. Their decision to live at Lumi, and to raise their children to school age there, was an unequivocal statement of their commitment to the people and the project, and so the local communities felt reassured about invest-ing their own time, effort and trust in the Thomases.

It's not only Papua New Guinea that faces the extinction of its larger mammals and other vulnerable species. Throughout the Pacific, but par-ticularly in the Solomon Islands, Fiji and New Caledonia, bird and mam-mal species stand on the brink of extinction, and without effective intervention a cascade of unique bats, rodents, pigeons, parrots, raptors and other vulnerable species will fall into the extinction pit. Further to the west, in Indonesia and Malaysia, a number of endangered species also cling to existence by the narrowest of margins, and none are more threat-ened than the region's two species of rhino.

Until 2010 the Javan rhino survived as two populations – one in Ujung Kulon reserve on far western Java, the other in the highlands of Vietnam. At a tonne and a half in weight, Javan rhinos are true mega-fauna – magnificent creatures, which a century or more ago could be seen from Myanmar and north Vietnam to eastern Java. Just two years ago poachers killed the last of the Vietnamese population, and today somewhere between thirty-five and forty-four individuals, just six or seven of which are thought to be fertile females, eke out an existence in western Java.

Java is one of the most densely populated islands on earth, but when the Dutch settled what is now Jakarta in 1607, Javan tigers (now extinct) and rhinos roamed the outskirts of their settlement. We owe the survival of Javan rhinos to Krakatoa: when the volcano erupted in 1883, it devastated the villages of the western tip of Java, leaving the area depopulated. Somehow a few rhinos survived the cataclysm, and the Dutch colonial administration proclaimed a nature reserve, Ujung Kulon, to protect them. Krakatoa, however, erupts frequently and with great violence, meaning that unless another population of Javan rhinos can be established elsewhere, the species is living on borrowed time.

You might think that, having been reduced to such a tiny population and with the threat of volcanic Armageddon looming, the Javan rhino is in a hopeless situation. But in fact other rhino species have recovered from similarly low population numbers. In the early twentieth century southern Africa's white rhino population was down to twenty or so, but today it stands at over 20,000, making it the most secure of the world's five rhino species. Moreover, there are areas of suitable habitat available in Java as well as elsewhere in Indonesia where the species could be reintroduced. What is needed to save the Javan rhino is a properly funded and well-formulated plan, for which progress is scientifically monitored and outcomes are reported upon.

At around half the weight of the Javan, the Sumatran rhino is the smallest living rhinoceros species. A century ago it had a range almost as wide as its Javan cousin, and even fifty years ago it could still be encountered in many parts of Malaysia, Sumatra and Borneo. That, however, was before the oil palm boom, which saw most of its habitat destroyed and any rhinos encroaching on the palm groves snared or shot. Today less than 200 Sumatran rhinos survive. Their extinction would be truly tragic, for the species is a living fossil whose bones are not much different from those of rhinos which lived around 20 million years old. Remarkably, its nearest relative is the extinct woolly rhinoceros of Siberia.

Over the past few decades forty Sumatran rhinos have been brought into captivity, but there's been very little success at breeding them. A glimmer of hope came in 2012 for this venerable species when, on 23 June, as the result of a globally coordinated program, a baby was born to a captive female held at the Way Kambas rhino centre in southern Sumatra. The initiative had been supported by an Australian not-for-profit organisation, the Asian Rhino Project (ARP), which was established by another zookeeper, Kerry Crosbie from Perth. The ARP raises funds and works on everything from snare removal in reserves to captive breeding programs.

The last Javan and Sumatran rhinos face many problems, but their most important challenges lie in areas where Australians have considerable expertise. For example, a fence is required to make the Ujung Kulon reserve secure from human and domestic animal incursion, and as we've seen, Australians are now experts at building fences that protect wildlife. The rhinos may also be suffering from the spread of the arenga palm, an inedible, invasive species, and from competition from banteng (a kind of wild cow), which are overabundant in the reserve. Because of the large number of weeds and feral species in Australia, Australian researchers are well versed in managing such threats.

I wrote part of this essay in Malaysia, on a short holiday. The island I was staying on looked idyllic, but each day the sky was so filled with smoke from the burning of the rainforests of Sumatra, which were being cleared to plant oil palm, that I couldn't see more than 500 metres across the water. And each morning great rafts of human refuse, including flotillas of rubbish-filled plastic bags, floated serenely by on the tide. Among the garbage were half-rotten fish discarded by the trawlers that seemed to be perpetually prowling the waters of the Malacca Strait. It's what the environment looks like in a crowded world when strong environmental regulation and species protection are lacking.

The pollution of the air and waters I saw off peninsular Malaysia does not respect national boundaries. Smoke and other particulate pollution in Asia may influence Australia's climate, while garbage from Malaysian waters can wash up on our beaches, choking turtles and fouling shorelines. Self-interest alone should see Australians taking a leading role in regional environmental protection. But for me there are other, more urgent reasons. What sort of world would we be creating if we let the Sumatran or Javan rhino slip silently into extinction? How poor would future generations be if theirs was a world stripped of rhinos, tree-kangaroos and other endangered species? Australians possess the know-how to help avoid such an outcome, and the cost of doing so is modest.

Australia is expanding its foreign aid budget by a billion dollars in coming years. If even $100 million – just one-tenth of that increase – went to a biodiversity protection fund, huge progress could be made towards protecting the region's most vulnerable species and showing the world that Australians believe natural diversity is a precious inheritance which we hold in trust.

THE NEED TO ACT

This essay has ranged widely, from scientific investigations to the deep history of the Australian environment, to the folly of ideologues and the incompetence of governments, and to our collective failure, in recent times, to protect our native species. Some may see this failure as a signal to give up or move on to other issues. But what we need to remember is that we know how to solve this problem. It is not like many complex economic and social issues, where key factors lie outside Australia's control. Furthermore, the costs are not great, and the expertise required is in place. Nothing is keeping us from success except our failure to be accountable – to ourselves and future generations. Quantify the problem, devise a plan to deal with it based on sound science, and report on the outcomes. And keep the politics out of it. That is the approach required if Australia's magnificent natural heritage is to flourish into the future.

How might this be achieved? A legislative commitment by Australians to prevent further extinctions – a policy of zero tolerance – would be a good start. That commitment could be backed up with the establishment of a Biodiversity Authority, an organisation independent of government but funded by it, with a clear mandate to invest in programs to prevent extinctions, both in Australia and overseas. Such an authority could be given unequivocal targets, and the failure to reach them should have strong consequences. Funding from such an authority could stimulate various institutions to work together to achieve desired goals. State-based national parks might begin managing biodiversity with not-for-profit organisations that have specific expertise in species conservation. Indigenous groups might begin working more closely with both parks services and others, as might fire authorities and graziers.

It's clear that we can protect threatened species much more effectively – and much more cost-effectively – than our current efforts suggest. To do so, however, we have to be willing to let non-government agencies take up work where they are expert, and trust them to play a role in securing the biodiversity of our national reserve system. If there was ever a time for Australian governments and people to work together on such a project, it is now. There are many reasons why Australia should be a leader on environmental protection: we live in a vastly diverse region – the only citizens of a developed country to do so, with a fine tradition of environmental research. And what we do will have a great impact on conservation regionally. All Australians – from both left and right – should take pride in the preservation of our nation's biodiversity.

The growing selfishness of some Australians, and the spurning of environmental values that goes with it, is a temporary madness for which future generations will despise us. The anti-patriotic hatred of anything labelled green that is currently infecting parts of far-right politics is particularly tragic. Conservation desperately needs the energy and capacity that markets, business and conservative thought can bring to bear on this great challenge. Moreover, the conservation of the full spectrum of living things is not a project for half a nation. It's a project for a planet. And Australia can play a vital role in realising it.

SOURCES

Professor Lesley Hughes (Macquarie University), Sarah Legge (AWC), Roger Beale (Climate Commission) and Kate Holden read this manuscript in draft and provided valuable comments. Jill Redwood (Environment East Gippsland), Peter Cosier (Wentworth Group), Hugh Possingham (University of Queensland), Michael Kennedy (Humane Society International), Atticus Fleming (AWC), Robert Hill and Alexandra Szalay contributed greatly to my understanding of various aspects of the issues discussed. Thanks, too, to Dr Tom Rich for his suggestions on the title. Finally, Chris Feik brought his usual editorial excellence to bear.

9   Australia's endangered species lists: Department of Sustainability, Environment, Water, Population and Communities, *EPBC Act Lists of Threatened Fauna and Flora*, www.environment.gov.au/cgi-bin/sprat/public/sprat.pl.

11   "each entity included on the list should have a detailed recovery plan": a list of recovery plans can be found at www.environment.gov.au/biodiversity/threat-ened/recovery-list-common.html.

12   "A review by the World Wildlife Fund (WWF) of conservation plans written for members of the kangaroo family": Michael Roache, *The Action Plan for Threatened Australian Macropods*, 2011–2021, WWF-Australia, 2011, www.wwf.org.au/news_resources/resource_library/?2940/The-action-plan-for-threatened-Australian-macropods-2011-2021.

12–13   "the woylie ... is listed as critically endangered": G.J. Yeatman and C.J. Groom, *National Recovery Plan for the Woylie*, Department of Environment and Conservation (WA), 2012, www.environment.gov.au/biodiversity/threatened/publications/recovery/bettongia-penicillata-ogilbyi.html.

17   "1966 classic and foundation text of Australian environmental conservation": Alan "Jock" Marshall (ed.), *The Great Extermination – A Guide to Anglo Australian Cupidity, Wickedness, and Waste*, Heinemann, Melbourne, 1966.

17   "A friend whom I'd known since my first long pants": Alan "Jock" Marshall, *Australia Limited*, Think – Or Be Damned series, Angus & Robertson, Sydney, 1942, p. 1.

18   "Australia ... is a third-rate country": Marshall, *Australia Limited*, p. 2.

18   "The Australian attitude" and "No sane man": Marshall, *Australia Limited*, p. 3.

19   "About one in every five or six hundred": Marshall, *Australia Limited*, p. 4.

19   "whatever wealthy Australians": Marshall, *Australia Limited*, p. 24.

19   "knows nothing about the flora and fauna": Marshall, *Australia Limited*, p. 34.

19   "It has always been strange to me": Marshall, *Australia Limited*, p. 61.

20 "Conservation was a hopeful": William Souder, *On a Farther Shore: The Life and Legacy of Rachel Carson*, Crown Publishing Group, New York, 2012, p. 264.

21 "The suggestion we have a duty": Chris Berg, "Schools might as well tell students who to vote for," the *Age*, 8 July 2012, www.theage.com.au/opinion/politics/schools-might-as-well-tell-students-who-to-vote-for-20120707-21o0i.html#ixzz20GzdPPtT.

22 "It was, after all, Victorian duck shooters": Marshall, *The Great Extermination*: p. 210. "I have said many harsh things about politicians": Marshall, *The Great Extermination*, p. 212.

22 "There is no objective reason why environmentalism": Souder, p. 336.

26 "only six species were listed as extinct": Marshall, *The Great Extermination*, p. 41.

28 "the idea was 'deeply flawed, often misleading, and short on crucial information'": Victorian National Parks Association, "Alpine cattle grazing – it's a park, not a paddock," vnpa.org.au/page/nature-conservation/parks-protection/alpine-cattle-grazing-%E2%80%93-it%E2%80%99s-a-park,-not-a-paddock.

30 "Despite the recovery team" and "Locking in twenty-year contracts": Bridie Smith, "Making himself extinct: 'absolute disgrace' prompts Leadbeater's possum scientist to quit," the *Age*, 12 September 2012, www.theage.com.au/national/making-himself-extinct-absolute-disgrace--prompts-leadbeaters-possum-scientist-to-quit-20120911-25qo5.html#ixzz27F8Zt3TQ.

31 "setting priorities in a rational and transparent way is crucial": L.N. Joseph, R.F. Maloney & H.P. Possingham, "Optimal allocation of resources among threatened species: a project prioritisation protocol," *Conservation Biology*, Vol. 23, No. 2, April 2009, pp. 328–38.

31 "90 per cent of the world's fisheries are overharvested": Laura Eadie & Caroline Hoisington, *Stocking Up: Securing Our Marine Economy*, Centre for Policy Development, September 2011, cpd.org.au/wp-content/uploads/2011/09/stocking-up_final_for-web.pdf.

32 "The public service cuts": Ben Cubby & Josephine Tovey, "Parks will go unstaffed as environment jobs slashed," the *Sydney Morning Herald*, 18 July 2012, www.smh.com.au/environment/conservation/parks-will-go-unstaffed-as-environment-jobs-slashed-20120717-228cn.html.

34 "the Newman government looks set to repeal": Rosanne Barrett, "LNP promises to develop Cape York as wild rivers laws buried," the *Australian*, 1 August 2012.

42–43 "This is where the heathland plants … evolved": for those interested in the evolutionary details, a brilliant overview of the way these environments took shape in response to a lack of nutrients and intense fires has been published:

G.H. Orians & A.V. Milewski, "Ecology of Australia: The effects of nutrient-poor soils and intense fires," *Biological Reviews*, Vol. 82, No. 3, 2007, pp. 393–423.

45   "In 2001 luminescence dating revealed": R.G. Roberts et al., "Archaeology and the Australian megafauna," *Science*, Vol. 294, 2001, technical comments 7a.

46   "In 2009 direct dating of teeth": Rainer Grün et al., "ESR and U-series analyses of faunal material from Cuddie Springs, NSW, Australia," *Quaternary Science Reviews*, Vol. 29, Nos 5–6, 2009, pp. 1–15.

46   "researchers investigated lake sediments in northeastern Queensland": Susan Rule et al., "The Aftermath of Megafaunal Extinction: Ecosystem Transformation in Pleistocene Australia," *Science*, Vol. 335, No. 6075, March 2012, pp. 1483–6.

48   "the largest and most distinctive warm-blooded predator": Kim Akerman & Tim Willing, "An ancient rock painting of a marsupial lion, *Thylacoleo carnifex*, from the Kimberley, Western Australia," *Antiquity*, Vol. 83, No. 319, March 2009, www.antiquity.ac.uk/projgall/akerman319.

52   "elephants should be introduced to the outback": David Bowman, "Conservation: Bring elephants to Australia?" *Nature*, Vol. 482, No. 7383, February 2012, www.nature.com/nature/journal/v482/n7383/full/482030a.html.

55   "fire served a variety of purposes": N.D. Burrows, A.A. Burbidge & P.J. Fuller, "Integrating Indigenous Knowledge of Wildland Fire and Western Technology to Conserve Biodiversity in an Australian Desert," www.maweb.org/documents/bridging/papers/burrows.neil.pdf.

57   "rapid decline of mammals in Kakadu National Park": J.C.Z. Woinarski et al., "The disappearing mammal fauna of northern Australia: context, cause, and response," *Conservation Letters*, Vol. 4, No. 3, June/July 2011, pp. 192–201.

58   Figure 1: from Woinarski, J.C.Z. et al., "Monitoring indicates rapid and severe decline of native small mammals in Kakadu National Park, northern Australia," *Wildlife Research* Vol. 37, No. 2, 2010, pp. 116–26.

61   "A study of 637 vertebrate species": J. Carwardine et al., "Prioritising threat management for biodiversity conservation," *Conservation Letters*, Vol. 5, No. 3, June 2012, pp. 196–204.

62   Figure 2: from Legge et al., "Rapid recovery of mammal fauna in the central Kimberley, northern Australia, following the removal of introduced herbivores," *Austral Ecology*, Vol. 36, No. 7, November 2011, pp. 791–9.

63   "As reported by the AWC in 2012": Australian Wildlife Conservancy, *Wildlife Matters*, Autumn 2012, www.australianwildlife.org/images/file/2012%20Autumn.pdf.

George Brandis

A surprising oversight in David Marr's essay on Tony Abbott is its failure to pay any significant attention to his time at Oxford. The "Oxford years" were decisive in the shaping of many Australian political leaders – Malcolm Fraser, Bob Hawke, Kim Beazley and Malcolm Turnbull among them. Fraser's visceral hatred of racism is often traced to the friends he made and teachers he encountered at Magdalen in the early 1950s, while Hawke's once-fashionable larrikin image has been endlessly burnished by tales of his heroic student drinking and similar laddish feats. But in Marr's account of Tony Abbott, although much is made of his time at the University of Sydney, a meagre paragraph is devoted to the two years, 1981 to 1983, that Abbott spent as a Rhodes Scholar at Queen's College, Oxford.

Marr's omission of Abbott's Oxford years is all the more surprising since it is clear from Abbott's semi-autobiographical book *Battlelines* that this was a crucial time in his intellectual development. He studied PPE (Philosophy, Politics and Economics) at the feet of many of the finest minds in the English-speaking world – for instance, his politics tutor was the great constitutional scholar Geoffrey Marshall. It was at Oxford that he met and developed an important friendship with the charismatic Jesuit intellectual Father Paul Mankowski. And it was during these years that his sense of vocation crystallised into a decision to enter the priesthood himself.

Tony Abbott and I were direct contemporaries at Oxford. We were members of different colleges, took different courses and moved in different circles. I would not say that we were friends. But in the small community of Australian students – at any given time, not more than a couple hundred – we knew each other well enough, and often ran into one another at functions put on by the Australian Society. To the best of my recollection, the first time I met Tony was in about November 1981. (The Oxford academic year begins in October.) The latest Dame Edna Everage show was on in London and a mutual Australian friend, Don Markwell, hired a bus and led an excursion of the newly arrived, slightly homesick Australians to enjoy an evening's entertainment by our gladioli-waving national icon.

I had quite a few encounters with Tony in the two years that followed. His staunchly conservative ideas were already firmly established; he was well to the right of me in those days. I was a member of the Liberal Party; at that stage Tony was not, but naturally we mostly talked – and argued – about Australian politics.

Tony Abbott has been called many things over the years, but I doubt he has ever been likened to Sebastian Flyte, the epicene, sexually ambiguous hero of *Brideshead Revisited*. Yet in one sense (and only one sense), Tony at Oxford was exactly like Evelyn Waugh's description of Sebastian: "he was the most conspic-uous man of his year." Everybody seemed to know Tony Abbott; at least, every-body had a story about him. The Poms, in particular, were fascinated by him, for he was everything they wanted Australians to be: authentic, gregarious, unselfconscious and engaging. While some of the twits looked down their noses at him, they were few. Among most students – men and women – Tony was immensely well liked; by the end of his two years, he had become not just one of the most recognisable but one of the most popular people in the university.

Tony has never been one to avoid controversy, and in Oxford it was no dif-ferent. One escapade that attracted attention occurred in April 1982, during the Falklands War. There are few places in the world more opinionated than univer-sity towns, and Oxford is more opinionated than most. Naturally, there were ardent supporters and opponents of Mrs Thatcher's decision to liberate the Falk-land Islands from the Argentinian invasion. In response to a major anti-Thatcher demonstration in the centre of the town, Tony took it upon himself to organise a pro-Thatcher rally. The fact that this had been organised by an Australian did not go unremarked. And, of course, there was his legendary success in the box-ing ring, when he won his Blue by flooring the Cambridge champion in an early round. This too contributed to his prestige – and, if I recall correctly, was a page-one story in the local newspaper, the *Oxford Times*.

In his Oxford years, many of the elements of Tony's complex personality were already on display. At one level, one saw the fearless controversialist who never backed away from an argument, and who indulged his Hemingwayesque taste in sports. But at a deeper level, a fascinating mind was being forged and a deep spirituality was being nurtured. I remember once when a few of us were engaged in that perennial late-night discussion among university students: what are you going to do when you graduate? Glittering careers in public life, academia or the professions were envisioned. When it was Tony's turn, he said simply, "I'm going to be a priest." Everybody laughed; we thought he was joking. His reputation was not of a person much given to piety. But he was adamant. That is what he had decided to do with his life.

Nobody feels more that the world is their oyster than someone graduating from one of the world's great universities, and Tony Abbott, with the accomplishments of a Rhodes Scholarship, an Oxford degree and a Blue to his name, confident and popular, had the world absolutely at his feet. But he had decided, after long reflection, to turn away from worldly things and give himself to the religious life of service.

Although, after three years at St Patrick's Manly Seminary, he eventually concluded that the celibate priestly life was not a discipline he could observe, the pastoral element has never left him. He remains devoted to the Jesuit ideal of a life lived for others – as his annual pilgrimages to live and work among Aboriginal Australians (unpublicised for years), and his lifelong commitment to voluntary service organisations, such as the Rural Fire Brigade and the Surf Lifesaving movement, attest. David Marr's conclusion, that for Abbott it has always been about power, could not be more wrong.

There is one particular memory of Tony Abbott at Oxford which sticks in my mind, as fresh as if it had happened yesterday, although it was thirty years ago. It was the summer vacation of 1982. I was heading off on a trip through Russia, and I knew that Tony had just been there. We ran into one another on the High Street one morning and chatted away. I remember saying how much I was looking forward to seeing the Hermitage Museum in Leningrad and various other cultural treasures. He looked at me sceptically. I asked him what he had thought of the Soviet Union. "Mate," came the reply, "mate, it doesn't have a single redeeming feature."

George Brandis

Chris Uhlmann

The Tony Abbott who emerges from David Marr's pen is a far more elusive figure than most of the discussion of the essay would suggest. In the months since it was published, enormous effort has gone into ensuring Abbott is reduced to one disputed paragraph: a dangerously angry young man who throws a punch at a wall near a woman's head. That impression of Abbott as an aggressive misogynist has been carved in stone by the prime minister's declaration of gender war.

But that Abbott is at odds with the Tony found elsewhere by Marr. We encounter a minister that staff and bureaucrats described as "admirably polite," someone who "never explored [cuts] with relish. People find this amazing but he doesn't seek conflict."

We learn he was devastated that cabinet overturned his "rock-solid, iron-clad" commitment not to lift the Medicare safety-net thresholds, and considered resigning. This fits with the memoirs of the former treasurer Peter Costello, who wrote, "Tony always saw himself as something of a romantic figure, a Don Quixote, ready to take on lost causes and fight for great principles. Never one to be held back by the financial consequences of decisions …"

On the signature Howard government policy WorkChoices, he is one of the few arguing against it, by saying, rightly, that it would undermine the battlers' faith in the prime minister. He warned cabinet: "It was always going to look as though we were exposing vulnerable people to danger."

This defence of the working man, and woman, rings with the voice of the disciple of B.A. Santamaria, deeply familiar with Pope Leo XIII's 1891 encyclical, Rerum Novarum. In it Leo examined the rights and duties of capital and labour and said:

> Let the working man and the employer make free agreements, and
> in particular let them agree freely as to the wages; nevertheless …

> If through necessity or fear of a worse evil the workman accepts
> harder conditions because an employer or contractor will afford
> him no better, he is made the victim of force and injustice.

Tony's Catholic voice is raised whenever he seeks to describe his political vocation. Aristotle and Thomas Aquinas echo when he tells Paul Kelly that politics is a way of giving glory to God: "This idea that politics is a managerial exercise, a simple question of resource allocation, I just think is dead wrong because politics is about inspiring people and persuading people there is value in what they do."

One of the church's key social tenets, "the common good," lurks in Tony's maiden speech, in his belief that governments must be an "instrument for giving social cohesion and purpose to our national life."

Intriguingly, during his first foray into parliament, when describing what government should do, he quoted Ben Chifley's "Light on the Hill" speech, itself drawn from an age when Labor was imbued with a deep understanding of the Catholic Church's social teaching. "People expect governments to work ..." Tony said, "'for the betterment of mankind, not just here but wherever we can lend a helping hand.'"

It's perfectly reasonable for Tony Abbott's political life to be informed by his Catholicism; many Greens are informed by a modern take on pantheism and no one seems troubled by that. And it's arguable that some of his best political impulses are those shaped by a rich tradition of theology and philosophy. It was therefore essential for Marr to examine the Opposition leader's faith, because it is impossible to understand Tony Abbott without it.

What should have been the talking point of the essay is the glaring fault-line Marr marks between Tony's faith and Abbott's ruthless pragmatism. Tony's better angels have ever been at war with Abbott's earthly ambitions, a tension that appears even in his decision to train for the priesthood at St Patrick's College, Manly. Why? "He wanted to be Archbishop of Sydney," Father Michael Kelly told Four Corners in 2010. Given the church leadership's immersion in the darker arts of politics for nearly 2000 years, Father Kelly might have added St John Chrysostom's observation that, "The road to hell is paved with the skulls of bishops."

Tony is guided by Christ's distillation of the law: "Do unto others as you would have them do unto you." Abbott is driven by a parody of that dictum coined by another Catholic politician, James Curley, the three-time Boston mayor: "Do others, or they will do you."

Tony used to emerge often in public but his appearances have been tragically curtailed since Abbott became Opposition leader. If Abbott now finds himself the victim of a campaign to reduce him to a caricature, then he should reflect on this: he has made it possible because he has spent a political lifetime reducing himself.

Marr notes that on the day after his by-election victory, Abbott told the press he was looking forward to being a "junkyard dog savaging the other side." He finds Abbott's colleagues "in awe of his savagery."

Marr says that in 2001, "Howard put the young minister's aggression to use by also appointing him Leader of the House." And this sentence is telling: "Abbott knew he had to counter the caricature of the Catholic hard man if his leadership ambitions were ever to go anywhere." Tony Abbott has known for a very long time that if he wants to be prime minister, he has to show the public more than his brute face. But he rarely has.

Each time an opportunity has presented itself for Tony to flourish, he has been overwhelmed by Abbott. As Opposition leader, Abbott has utterly suffocated Tony. Everything about him has been diminished, even the points of attack on the government. He has now taken as gospel the words of Lord Randolph Churchill: "Oppose everything, suggest nothing and turf the Government out." The contemporary take on Churchill's dictum was best distilled by one of Abbott's staff: "the job in Opposition is to get your boots on and kick the shit out of them." Abbott and his staff are convinced that Oppositions do not win elections – governments lose them. So the Opposition leader's role is to do as much damage to the Gillard government as possible.

This is a political take on an ancient theological approach, the via negativa: Abbott defines himself by what he is not. Above all he is Not Labor. There is a crude genius to this. On one level it has been an outstanding success. Gauged only by how much damage he has inflicted on Labor, Abbott is the most successful Opposition leader in Australian political history. He has undone one prime minister and may bring down a second.

But this scorched-earth politics ignores something crucial: how much damage Abbott has done to himself. His approach is consuming the better parts of his nature and defining him politically and personally. An attack on your character cannot stick unless it is plausible. The caricature of Abbott as a brutal misogynist is not accurate, but he has ensured that it is credible. The Coalition is clearly deeply worried that the caricature is now setting like cement in the public's mind. If it wasn't, it would not have wheeled out his wife, Margie, and his daughters to mount a defence.

And to date, Abbott has done the Gillard government great harm without becoming prime minister himself. It will all come to nothing if he does not succeed in taking that last step.

Perhaps Abbott should ponder what became of his political hero, Santamaria, and what the Movement did to his church. In Edmund Campion's brilliant reflection on growing up Catholic in Australia, *Rockchoppers*, he writes that the saddest moment of Santamaria's life seemed to Campion to be a dinner for the fortieth anniversary of the Movement in July 1981:

> B.A. Santamaria, the man who, all those years ago, had set out with a great ache in his heart to remake the world so that the poor and rejected could find compassion and justice, was ending his days among the stone-faced men of the Right.

And the church? Before the Movement there was a longstanding Catholic critique of Western society that:

> tried to determine what had happened to humankind in Europe and its dependencies since the Reformation and the industrial revolution; then it tried to assess this in terms of Christian humanist philosophy ...
>
> Nothing was outside the scope of this criticism – the factory system, universities, the press, monopolies, industrial farming, the control of capital, modern warfare, communism and socialism, restrictive immigration – it was assumed that an authentic Christianity had something sensible to say about all aspects of life.

But in the wake of the Movement, "Catholics got so badly burned they abandoned social criticism of this deeply philosophical kind." The way the Movement went about using its political power not only split Labor, it also split and damaged the church.

In the end Abbott might triumph no matter what Labor throws at him. Not being Julia Gillard could be enough to get him there, given the equally visceral reaction to the prime minister among large swathes of the electorate.

And right now, Tony Abbott might console himself with the idea that when he finally grasps the prize, he will govern differently to the way that he won government. But, in dealing with a hostile Senate, he might find that, as Paul noted: "Do not be deceived: God cannot be mocked. A man reaps what he sows." And as he wrestles, too late, with the task of convincing a sceptical public that there is more to him than meets the eye, he might reflect on the words of Luke's Gospel: "For what does it profit a man if he gains the whole world and loses himself?"

Chris Uhlmann

Mark Latham

Lost in the controversy about "the punch" in David Marr's Quarterly Essay on Tony Abbott was a more significant insight into the Opposition leader's character. I first met Abbott thirty years ago, during his notorious, rampaging period as a student politician. We were elected to the House of Representatives within months of each other in 1994, commencing a series of robust exchanges. Yet in this time, I never really got a handle on Abbott.

He was an enduring contradiction: someone who preached conservative values yet practised an aggressive, scandal-prone brand of politics. It was easy to write him off as a Mad Monk, a sobriquet his parliamentary colleagues readily embraced.

This is where the Marr essay is invaluable. It identifies the character trait which, above all others, explains the Abbott puzzle. He is a man of chronic hyperbole, an attention-seeker who cannot engage in public debate without exaggerating the faults of his opponents and their policy positions.

Marr sets out the trail of overstated behaviour. The pampered childhood, in which no boundaries were placed on young Tony's adventurism. His time at Sydney University hectoring lesbians and vandalising public property in the name of conservatism. His struggles as a trainee priest in conforming to the vows of celibacy and the culture of St Patrick's seminary. Then his turbulent period in the early 1990s, ostensibly working for the Liberal leader, John Hewson, but acting as an agent for John Howard.

His old boss has neatly defined the Abbott technique. "He gets right in your face," says Hewson. "He exaggerates, he grabs the headlines, even if he knows that the next day he's gonna have to back off." As a member of parliament, Abbott has been a habitual exaggerator. He has compared himself to Jesus Christ and Winston Churchill. He went over the top in his pursuit of Pauline Hanson and Cheryl Kernot – in the former case, admitting to lying about the funding of the anti-Hanson campaign. During the 2007 election campaign, he vilified the anti-asbestos campaigner Bernie Banton.

As Opposition leader, Abbott has become well-known for his fitness regime, typically at the extreme end of endurance sports. He is not the sort of person who runs around the block or swims a few laps to stay in shape. He organises 24-hour marathons and seven-day bike rides. So too, in his community service, he does not staff the school canteen or deliver meals-on-wheels. Instead, his publicity machine promotes his "action man" exploits: fighting dangerous bushfires and battling treacherous surf conditions as a life-saver. In his private life, Abbott does not pursue the safe and sensible; he enjoys an exaggerated sense of danger.

His stance on policy issues is no different. On the carbon tax, he could not help himself, running the most fraudulent scare campaign in the nation's history. His predictions of economic ruin now look ridiculous. Even when Abbott backs down, he uses embroidered rhetoric – on this issue, switching from a cobra strike to a python squeeze.

In the month since Marr's essay, the exaggerations have continued. George Megalogenis from the *Australian* has picked apart the amplified claims in Abbott's defence policy speech to the RSL. When the prime minister visited New York recently, Abbott overreached in telling her to go to Jakarta to talk to the Indonesian president, when Yudhoyono was already on his way to the Big Apple. Even in his wife's public appearances in early October, the Opposition leader went for over-kill, using six media events when one would have been more appropriate (and less risky, especially if the "Good Man" endorsement rebounds on him).

This pattern of hyperbole is the reason why the electorate has never warmed to Abbott. Sure, he has been around a long while and people know a lot about him. But the public has little sense of who he is. Conservatives are supposed to be moderate, careful, temperamentally cautious types, not tearaways forever shooting off their mouths. People think there is something wrong with Abbott: he talks about a particular set of beliefs in a way which undermines the integrity of those values.

Under his leadership, we are witnessing the moral decline of Australian conservatism. Howard set a credible public standard because, in large part, his personal style matched his policy ideas. Under Abbott, the Liberal Party's methods and values have become disconnected. Reckless exaggerations are undermining their message about prudence and traditionalism in public life.

Naturally, a leader's words set the tone for contributions by his followers. It is not by coincidence that Abbott's close friends Alan Jones and Cory Bernardi have caused him grief recently with their wild, irrational claims. Incapable of backing away, Abbott resurrected Jones's "died of shame" phrase in parliamentary debate with Julia Gillard on 9 October – a recklessness his deputy Julie Bishop could not defend when she appeared on Channel Ten's *The Project* later that day.

I have no doubt Abbott's words, in supporting his motion to remove the Speaker in the House of Representatives, were carefully planned and rehearsed, as are each of his major set-piece speeches in parliament. His excuse that he had forgotten Jones's attack on the prime minister's deceased father is not credible – an echo of his alibi in Marr's essay that he had no recollection of the Barbara Ramjan punch. Abbott should carry this slur against the Gillard family as a ball-and-chain on his character and credibility for the remainder of his time in public life.

Some critics have described Marr's essay as the politics of personality, a spark for the inferno of character attacks now dominating the Australian parliament. History should record, however, that the first venture into personality politics in this term of parliament was launched by the *Australian* newspaper in its allegations against Gillard from her time as a lawyer at Slater & Gordon in the early 1990s. The prime minister's character having been publicly tested in this fashion, Marr's essay gained extra media attention as a legitimate assessment of Abbott's make-up.

This is a debate Labor cannot lose. Abbott's hyperbole habit is so entrenched that his best chance of winning the next election is on policy issues, not personality. Indeed, this is the break in the weather the government has been waiting for. In a test of character and temperament, Gillard is making ground against a chronically flawed opponent.

Mark Latham

A version of this comment appeared in the Australian Financial Review *on 11 October 2012.*

Judith Brett

In *Political Animal*, David Marr gives us two Abbotts: Politics Abbott and Values Abbott. The first is the man of driving ambition who competes hard for the prizes of political office; the second the man who is a self-conscious conservative, defender of the monarchy, the faith and the family, and motivated by his deep religious beliefs. Marr shows the shifting tension between these two Abbotts, and argues that when they are in real competition with each other, it is Politics Abbott that matters – the aggressive, competitive, ambitious guy who wants the top job no matter what – and that his values have never stood in his way. Maybe, but he is one man, and I think we need to try to connect the two Abbotts more closely than this, to understand the interdependence of Politics Abbott and Values Abbott. This is a tough call and what follows is not a fully worked-out answer, simply some thoughts which might point the way.

Much has been written about Abbott's problem with women, that he prefers them in their traditional family roles of wife and mother and has conservative views about sexuality. In holding such views Abbott is out of his generation, a post-feminist man who has the attitudes of men born before World War II, the type of men in fact who became his mentors. To my mind these attitudes do not make him a misogynist. Abbott's rather patronising and outmoded preference for women in their traditional roles is not the same as a fear and hatred of women. Abbott seems to like women, to enjoy their company, as did many men in preceding generations. What is prima facie puzzling, however, is how and why Abbott has resisted the profound changes in gender roles of the past fifty years. Yes, he grounds his attitudes in his religious faith, but even so there is something of the contrarian here, as there is in his passionate defence of the monarchy. Why does Abbott need to stand out against the crowd and the times, to identify so strongly with the men of the past that he puts barriers between himself and so many of his own and younger generations of Australians?

Whatever the answer, Abbott's attitude to family, gender and sexuality is only part of the problem that I, as a woman, have with him. A bigger problem for me is his hyper, compulsive masculinity and a relentless competitiveness which robs him of judgment. Whatever was he thinking when he accused the dying anti-asbestos campaigner Bernie Banton – who was protesting about the cost of an expensive drug to treat mesothelioma – of pulling a stunt and thinking he could get away with it because he was sick? Bernie's problem was that protesting against the government put him on the other side, and so made him fair game psychologically for Abbott's oppositional politics and his angry tongue. Although Abbott enjoyed boxing in his youth, it is verbal not physical aggression which is his problem. He is not like Mark Latham, whose crushing handshake of John Howard reminded voters of his assault on the taxi driver; and whether or not Abbott did punch the wall on either side of Barbara Ramjan's head, he did not hit her.

Abbott's problem with verbal aggression is most on display in the theatre of parliament. In its origins parliament is about the taming of aggression, providing a space in which men fight with words rather than with swords and fists, resolving political conflict and rival ambitions through the elaborate rituals of organised debate. Conflict is over both policies and power, and Abbott's problem for many of us is that in his relentless, negative opposition to everything the government does, he seems to be only about power. The other problem is that he clearly can't stand it that he is not the prime minister. The presence of women in Westminster parliaments organised around bipolar rivalry for office is profoundly destabilising. Abbott seems absolutely furious that it is Gillard and not he who is prime minister. I am not sure if he would be any less hostile and rude if the government were led by a man, but with it being led by a woman the rules of combat are less clear, the boundary between legitimate public arguments and illegitimate attacks on aspects of his opponent's private life harder for him to negotiate. Hence his undisciplined, underhand and unacceptable digs at Gillard's marital status. Rivalry is a major source of aggression, and when the rival is a woman the aggression cannot help but look excessive.

Marr discusses Abbott's belief that governments can give cohesion and purpose to our national life, and points to his sensitivity to the fragility of society, his conservative fear that too much change will unleash emotions that society will find hard to control. Conservative pessimism about the unanticipated destructive consequences of radical change versus progressive optimism about its creative and liberating potential has a long history, but it also has a psychological dimension. Those who are closer to their own destructive impulses and emotions are more aware of the value of traditional social controls, of law and order, in holding impulse in check. Abbott is not really a law-and-order man, which anyway is less relevant in federal politics; but he does give his aggression plenty of space, and so is more aware than those of milder temperaments of the importance of rules and institutions for containing it. Perhaps this is the link between Values Abbott and Politics Abbott.

This still this leaves some big questions, however. Why is Abbott so angry? And if he defeated his rival and won the prize, what would then drive him? What is an angry man like once he has become prime minister? Would we then see the triumph of Values Abbott, Politics Abbott having cleared the way for him to do the good he always intended? Or is he hardwired for rivalry and anger?

Judith Brett

Jack Waterford

In the standard hagiographies, Ignatius of Loyola was a soldier, good-time Charlie and libertine who while wounded read words of Jesus (Mark 8:36) asking: "For what shall it profit a man, if he shall gain the whole world, and lose his own soul?" Ignatius underwent conversion and formed the Society of Jesus, or Jesuits, a religious order with particular fervour for defence of the Catholic Church against the arguments of the Reformation, for education and intellectual rigour, and loyalty to the Pope.

At Jesuit schools, the phrase that challenged Ignatius and the questions it invites about the purposes of power are drilled into students. Yet Jesuits have always been accused of ever being worldly, as focused on preparing their sons for leadership and power in this world as in the next. John Howard, as prime minister, saw a big transition of Catholics into the Liberal Party, and into the ministry. Almost all were taught at Jesuit schools.

It was quite early on that Dick Abbott knew that he and his wife, Fay, had produced something out of the box in their infant Tony. His mum, apparently, was given to telling others that he would one day be Pope or prime minister. Women in the family – mother and sisters – favoured the latter. Bob Hawke's mother had a similar annunciation, a reason for not discounting such revelations.

Abbott went to Saint Ignatius' College at Riverview, where a famously urbane Jesuit, Emmet Costello, also saw his leadership qualities. Costello told David Marr that "from the moment I met him, he was different. He walked into my room – I was chaplain for the boys – and he projected an image immediately of high intelligence, ambition, drive and leadership, and I thought this guy is worth following." Costello, whom Abbott rated second to his own father as the most important influence on his life, thought Abbott might become a priest.

David Marr's essay on Abbott sees much of the character in Abbott as the tension between that would-be priest and the would-be politician. He tried the seminary and decided it was not for him. But before and after, he has been known for fierce and conservative moral opinions, particularly on sexuality and abortion, with critics suggesting he is a fanatic and prig, seeking power so he can ram his opinion down the throat of others.

Even Marr has his worries on this, though he recognises what many others do not – that while Abbott will not abandon or disown his values or principles, he will not be a martyr to them either. In a way which might not parse comfortably with Loyola, he does not see power or the world simply for spiritual satisfaction. The world itself is not for redeeming.

Marr contrasts the would-be priest (Values Abbott) with the would-be prime minister (Politics Abbott). The latter, Marr is quite sure, is the one that matters.

> That's the one who got him where he is today … His values have never stood in his way. In the past he has talked about being prime minister to make a difference, to allow Australians to be their better selves and – again and again and again – to ensure a more cohesive society … He's a worker. No doubt about that. But the point of it all is power. Without power it's been a waste of time.

That's a bleak picture, albeit from a critic unlikely ever to have much warmed to Abbott. But it is a criticism which can be made of many other politicians. It's hard to see what Julia Gillard wants, and if one ever had a strong sense of where John Howard, or Bob Hawke, stood or where they were headed, one does not see them as being on a crusade towards any particular light on a hill.

It was said of Paul Keating that he became so exhausted by the manoeuvring, backstabbing, brawling and deals of his rise to the top, and the assassination of his predecessor, that he lay a long time stunned when he got there, as if mere achievement of the top had been what it was all about.

Each political party and each politician struggles between idealism and realism. Effective politicians understand power is the first and the last thing, and that one cannot achieve anything without it. Justifying dodgy means by the nobility of the ends is routine. That involves compromise, ambush, deceit, often hiding one's ambitions, often going in what one believes to be the wrong direction. It often means swallowing pride and biding one's time. And betrayal of friends.

Getting power and holding it is the absolute prerequisite for exercising power. Some want to be pure, regard any compromise as betrayal, and would rather lose than surrender on any principle. Gough Whitlam mocked this in the ALP Left in the 1960s when he told them that "only the impotent are pure."

But sharp as Marr is, my own feeling is that the real tension in Abbott lies rather more in Christ's words than between his would-be priest and would-be king. He's made, in effect, a Faustian deal, with his colleagues at least, which now cripples not so much his efforts to get power, but to exercise it, when it comes, in any instinctive, or moral, way. That's in part because he does not really belong in the Liberal Party – and that he knows it, and that many of his colleagues know it and thus have a leash on the power they will let him exercise. It's in part because his political tactics have been amazingly successful, but that his strategy is not so well appreciated.

His colleagues admire the amazing self-discipline over three years, by which Abbott has not been himself – so little himself indeed that the real Tony may have actually disappeared. That will probably put them in power.

But gratitude will not extend to letting him revert to his sheep's clothes, to giving him his head, or the keys to the till. And if karma, to mix the religious metaphor, has anything to do with his future, he may also have already made the choices that will be the ultimate cancer on his authority.

Far from being a dangerous radical, or a fanatic with a secret agenda (whether to restore WorkChoices or punish abortion, for example), the risk may be that Abbott will be both timid and timorous as prime minister. Condemned, more or less, to imitate one other man who made him a protégé, John Howard, in calm but essentially reflexive management of events, rather than pursuit of any particular mission. (That's a bit unfair to Howard, who could move abruptly and very laterally when politics required it, but Abbott himself has never shown either the footwork or the supreme political judgment of a Howard. In football, Abbott was a front rower, not a half-back.)

Voters have had long enough to see the Abbott temperament, and much of his character. The idea that he is holding back or hiding something is probably wrong.

But equally wrong might be seeing him as a person still hiding promise or potential. I expect that what one will get – for good or ill, very much a matter of personal judgment – is pretty much what is now on view. The risk, for those whose inclination to vote for him comes primarily from anger at or rejection of Labor, is whether not being Labor, or Gillard, is enough.

Jack Waterford

This comment appeared in the Canberra Times on 12 September 2012.

David Marr

I shouldn't have been surprised the punch hit so hard. Abbott and his troops had been stumbling over women for weeks. After sledging one, the prime minister, and defying another, the deputy Speaker Anna Burke, Abbott became the first leader of the Opposition for a quarter of a century to be thrown out of the House. His woman problem was back on the table. And the Liberals kept stumbling. Party strategist Grahame Morris called 7.30's Leigh Sales "a real cow" and apologised. "Women are destroying the joint," declared the radio godfather of the party Alan Jones. He didn't apologise. Gillard lashed out at "the misogynists and the nut jobs" pursuing her on the internet. Anne Summers documented their vilification, on and off the net, in a lecture widely republished in these weeks. The release of RU486 to all the doctors of Australia revived memories of Abbott's last-ditch stand against the morning-after pill. Gillard took the opportunity to assure women her government would always protect abortion rights. All through these weeks, the prime minister's stocks were rising. On 8 September, the Quarterly Essay appeared and with it Barbara Ramjan's account of the night in 1977 when Tony Abbott lost a student election: "He came up to within an inch of my nose and punched the wall on either side of my head."

The punch became the essay. Labor declared Abbott a misogynist bully. Gillard's front bench made it seem the leader of the Opposition was still getting about with bloodied knuckles. Abbott went to ground. No one could remember a time he had been so inaccessible. There were no doorstops and no visits to factories crippled by the carbon tax. But Abbott's friends sprang to his defence. None were more fiercely protective than other old followers of Bob Santamaria who now command the conservative heights of newspaper commentary. Gerard Henderson in the Sydney Morning Herald did Abbott no good by sneering at Ramjan's statement as "the uncorroborated testimony of the aggrieved." That provoked devastating retaliation from the Sydney barrister David Patch:

I did not see the incident, but I was nearby. The count had just finished. Ramjan found me. She is a small woman, and Abbott was
(and is) a strong man. She was very shaken, scared and angry. She
told me Abbott had come up to her, put his face in her face, and
punched the wall on either side of her head. So, I am a witness. Her
immediate complaint to me about what Abbott had just done had
the absolute ring of truth about it. I believed Ramjan at the time,
and still do.

This seemed to send Greg Sheridan over the edge. Another of the Santamaria
alumni and foreign editor of the *Australian*, Sheridan had declared the punch
"utterly inconceivable" and its appearance in the essay proof of my "overall slop-
piness as a journalist, failure as a historian and distorting bias as a polemicist."
The morning after Patch emerged, Sheridan lost his composure on ABC News
Radio. With Marius Benson barely able to get a word in edgewise, he raged for
ten minutes against Patch, me and the ABC for conniving in a "disgraceful, sec-
tarian, anti-Catholic campaign" against Tony Abbott. The climax was quite
something:

> SHERIDAN: Why would the ABC give a worshipful, uncritical and
> unbalanced interview to David Marr on *Lateline* whose article is full
> of inaccuracies, full of inaccuracies, when Marr is a committed man
> of the left and motivated by an obsessive hostility to Catholicism
> and not even put someone else on a program to balance him –
> BENSON: I think if we keep going along those lines we'll have to
> have David Marr with a right of reply so I'll leave it there, Greg
> Sheridan –
> SHERIDAN: Well, why wasn't there a right of reply to David Marr
> on *Lateline* when the ABC ran its sectarian, anti-Catholic –
> BENSON: Well, people have been –
> SHERIDAN: The ABC should be utterly ashamed. They would not do
> this to any other politician and it is a disgraceful episode and I'll tell
> you what, people pay taxes to the ABC who are also Catholics and
> who also vote Liberal and they shouldn't be forced to do so when the
> ABC performs in this way as an agitprop agent of sectarian hostility.

Abbott broke his week-long silence on Nine's *Today* show. Despite Patch and
another witness – unnamed but willing to go on the record if need be – Abbott

was still insisting to Karl Stefanovic that the punch never happened: "There is no doubt that the Labor Party dirt unit is running a serious campaign here and I've been saying to Margie and the kids, Karl, that they can expect a lot more of this between now and polling day." Abbott was playing the victim. A few hours later on the steps of an IGA supermarket in the Canberra suburbs, he repeated his "dirt unit" line. He offered nothing to back the charge.

I was flying up from Melbourne that morning unaware of Abbott's smear and unaware my publishers had rushed out a statement denying I'd been fed the story by a Labor dirt unit. (I'd first heard about the punch early this year from a number of judges at the fortieth anniversary of my law school graduation class.) As I walked in the door at home, my phone rang. It was Abbott. The conversation was not off the record. I took notes. "He said he was not accusing me of being part of any Labor Party dirt unit or operation against him and that he was taking care not to blame me." I thanked him for that. He added that the essay was thoughtful and intelligent. "While I don't by any means accept all your judgments, you did a highly professional job." Only after we spoke did I have a chance to go online and read the transcript of the IGA doorstop. I saw why he'd rung. He had dropped me right in it:

> QUESTION: Mr Abbott, are you suggesting that the Labor dirt unit is feeding this information to David Marr, this specific allegation? Is that what you're saying today?
> TONY ABBOTT: There is a Labor dirt unit and it's feeding information to people left, right and centre.

Ramjan and Patch put out their own statements that day denying the dirt unit allegation. Ramjan is a respected figure in Sydney, with connections to the highest levels of the judiciary. But the shoring up of Abbott's denials required the destruction of her credibility. Michael Kroger took on the task armed with a leaflet Ramjan had written during the student brawls of the late 1970s accusing Spartacists of threatening to kill her. His first stop was the Andrew Bolt show, where he accused Ramjan of lying about the punch and lying about the death threats. He also gave the leaflet to the *Australian*, which next day reported on page one:

> Liberal powerbroker Michael Kroger yesterday accused the woman who levelled allegations of physical aggression against Tony Abbott during his student years of being a serial manufacturer of false complaints against her political opponents.

That night Kroger was on Steve Price's radio show in Sydney calling Ramjan "an ex-communist who is now a nobody, a nobody attacking Tony Abbott to get publicity" and "lying and cheating against Tony Abbott in a fairly disgusting way." Next morning he was with Alan Jones calling Ramjan "a nutter from the left" involved in "one of the most sinister, nasty and vicious campaigns I've seen." Their grim exchange was laced with effusive praise for Abbott.

Ramjan and Jones have as a mutual friend the colourful Sydney lawyer Chris Murphy. He spoke to the broadcaster. Next day Jones made an apology on air as complete and elegant as is imaginable: "Now, on this program we try to be fair no matter who is involved. Since that interview yesterday it has been pointed out to me by several people that Barbara Ramjan is indeed a woman of distinction who in whatever she says deserves to be taken seriously ..." As this response goes to press, Ramjan is suing Kroger and the *Australian*.

How Abbott handled the punch says a good deal about the politician he is and the prime minister he might be. It was handled terribly badly. He might have admitted those blows and said, truthfully, that he has done a lot of growing up since 1977. He might have kept to the safe ground of not remembering what happened that night. Instead he shifted to denying the punch absolutely. That required attack dogs to be unleashed on Patch and Ramjan. He might have faced the press early, but allowed the story to veer out of control as he bunkered down day after day. He might have refrained from making the "dirt unit" allegations which quickly proved an invention. In that week Abbott showed himself to be furtive, aggressive, untrustworthy and, yes, not so great at dealing with women. Somehow or other he had managed to turn an ugly little incident from a long time ago into a big deal in 2012.

Misogyny was the word on every politician's lips. Two teams of women – Labor's handbag hit squad and the Coalition's twinset vigilantes – fought over Abbott's reputation with women in life and politics. His mother, his wife, his sisters and his daughters were produced for the cameras, an operation made all the more mawkish by the revelation that Abbott would rather watch *Downton Abbey* than the football. In interviews all through this I kept saying I didn't think Abbott was a misogynist. At heart he is an old-fashioned Catholic sexist but he doesn't hate women. He's not incapable of retraining. I look back, like Clive of India, astonished at my own moderation. Into the skirmishes over Abbott and women in these weeks fed the scandals of Alan Jones and Peter Slipper. The broadcaster was in disgrace for saying Gillard's father died of shame because of the prime minister's lies. What was left of the Speaker's reputation was trashed when it emerged that he had likened women's genitals to "a mussel removed from its shell." In a debate over Slipper's fate Abbott managed to conflate both scandals by accusing Gillard of leading a government already dying of shame. It was a devastating slip of the tongue from the old Abbott, not the highly disciplined Abbott who has been leading the Opposition for the best part of three years. And it provoked from Gillard a white-hot rebuke that has reverberated around the world: "I will not be lectured about sexism and misogyny by this man. I will not ..."

The numbers are going the wrong way for Abbott. While I was working on the essay in July, he seemed not far from invincible. The Nielsen poll gave the Coalition a twelve-point lead over Labor in the two-party-preferred stakes. Their poll the other day showed that lead cut to four points. In July, Abbott was the preferred prime minister by five points. Now Gillard is preferred by ten. Her satisfaction rating was minus nineteen then and is minus one now. His was minus seventeen then and is minus twenty-three now. How much the gender wars played into that poor result for Abbott is far from clear. Gillard's standing rose after the misogyny speech and rose particularly sharply with men. Nielsen pollster John Stirton puts the drift from Abbott down to the collapse of the panic over the carbon tax. That's showing up, too, in the focus groups run by Ipsos Australia. "The carbon tax isn't coming up at all," Rebecca Huntley told me. "People aren't worried about the tax now, certainly compared to how they were feeling in September of last year. There is no push to have it abolished either." They aren't too keen on Abbott. "There is an element of exasperation and despair about the debate because the government is disliked and so is the leader of the Opposition. There is a strong sense Labor will lose the next election and that it's worrying that this guy will be our prime minister."

But the election is still the best part of a year away. That's long enough for Labor to waste the opportunities the Coalition is presenting Gillard's beleaguered government. And even now it may not be too late for Abbott, if he has the inclination, to make himself respected, even prime ministerial. As I wrote in the essay: a year is an eternity in politics.

David Marr

Rachel Nolan

Laura Tingle's *Great Expectations* captures a critical and little-discussed trend in the national mood. Having spent eleven years in parliament, I thought it worth putting the perspective of someone who has been inside the vortex of that anger.

I strongly agree with two points in the essay: that Australians have little philosophical grasp of the (rightful) diminution of governmental power which deregulation has brought; and that we are angry at government without having a coherent view of what we want from it.

From 2009 until 2011 I was the Queensland Minister for Transport, a role that has a different significance in that state than in others because the railway is so central to Queensland's social history. When I became minister, Queensland had by far the largest state-owned railway in the country, with 15,000 staff and more than 10,000 kilometres of track. Queensland Rail Limited had been corporatised in the mid-1990s and was supposed to be run commercially by a management answerable to its board. As minister I had a narrow power to give directions to the board, but by convention this power was very rarely exercised.

Nonetheless, in my first year as transport minister I had a different rent-seeker at my door every day of the week. The mining industry said it couldn't afford Queensland Rail's shift to the more commercial "take or pay" contracts, the beef barons didn't want to be bound by real cost pricing – or indeed by any kind of contract at all – and the unions came to me because they were unable to negotiate with Queensland Rail on even the most basic industrial issues. I once found myself being heavied to intervene on whether or not workers at the Rockhampton Railway Workshops should be allowed to wear shorts. Seriously.

In all cases the rent-seekers – unable to get what they wanted through the corporatised structure – threatened that if they didn't get it from me, they would run an anti-government political campaign. And they did.

The ploy worked because even though Australian parliaments under Keating's National Competition Policy had shifted the country to corporatised models and real cost pricing for railways, electricity, water, telecommunications, forests and ports, no one really bought the idea that government had diminished its own power. Everyone believed that – corporatised structure or not – while government still owned the company, it was government that was accountable.

The current cost-of-living debates about electricity and water – which in most jurisdictions are run by state governments under a corporatised model – are more examples of the same thinking. In principle, everyone accepts that, except for the poor, households and businesses should pay the real cost of utilities. But when they are government-owned and bills are going up, the whole country will, as it is doing right now, emit a loud collective moan.

Tellingly, in the second year that I was transport minister, our government was fully committed to rail privatisation and the rent-seeking magically stopped. Although the sale had not yet gone through, the sectional interests now understood that we were serious about running the show commercially – and in the context of an almighty row over privatisation, a campaign on freight pricing wasn't going to hurt the government one bit.

When the privatisation occurred, no one lost a thing to which they were properly entitled. Miners have no reason to be sponging off the public balance sheet for purely commercial rail infrastructure; the export beef industry is not entitled to a hidden subsidy of $20 million a year; and there's no public interest in railway workshops and head offices being overstaffed at taxpayers' expense.

Australia made those decisions fifteen years ago with National Competition Policy, but by keeping the assets in public hands we kept governments politically tied to them – and therefore to the political priorities of yesterday.

Governments need to let go of commercial assets in order to focus on the right priorities for a new Australia. It's also true that governments need to privatise to free up cash for infrastructure, but for me that's not the main point. Australians must decide what we actually want our governments to do. I posit three things:

- provide genuine public goods such as defence, and law and order; public transport infrastructure; and social services – disability support, health and education;
- support innovation in infant industries; and
- meet the challenge of sustainability.

Governments, of course, must provide social services. The American experience – where first-time Australian visitors are uniformly shocked by the sight of the homeless poor begging on the street – shows us what you get in a society that can't be bothered to pay for basic levels of care. And while there's room for community and private sector innovation in delivering social services, if a comprehensive safety net is to be maintained, it must remain a core government responsibility. The cost of this, as shown by the National Disability Insurance Scheme and health costs that continue to grow at nearly 10% a year, will only go up. In a country unwilling to adopt European rates of taxation, governments simply have to dispense with some old responsibilities in order to keep up our standard of care.

Public transport and better road networks make for better cities (where, despite the nonsense mythology of rugged outback life, most of us live), and, as a series of failed public–private tollway partnerships have shown, only governments can afford to deliver these.

Our national economy will have to innovate to flourish in an increasingly globalised world. A fleet-footed government would channel its limited industry assistance towards bright ideas – some of which will fail – rather than into mature or declining sectors, like agriculture and cars, which only get the cash because their workforces have political power derived from their concentration in a handful of federal electorates.

And of course the sustainability challenge is so big that only government can lead.

These three things – social services, innovation and sustainability – are, in my view, the core responsibilities of Australian governments. They are modern, they are progressive and they are simple ideas in which people can discern the clear purpose of government. We should do everything else for ourselves.

Fifteen years ago, under the Hawke and Keating governments, Australians began the process of reimagining what our governments need to do. But the failure of the Howard government to follow through those reforms means we now find ourselves in a philosophical malaise.

Conduct a focus group in any part of Australia right now and you'll be told that the cost of living is the hot political issue. It's been a key campaign topic in every election since the Rudd win in 2007, and yet, as public policy issues go, it embodies a nonsense. While state Oppositions (of either stripe) criticise state governments for price increases, no one really thinks it's possible to make the basics much cheaper – and only a bureaucratic lunatic would think it's efficient to subsidise power or water costs from the consolidated fund.

Rather, governments are stuck in a no-win political argument over something which they cannot control, and indeed which they quite simply shouldn't own. And being stuck in that place means they are mired in a political quagmire of their own creation. You cannot run the country or set it on a progressive path while you're still in a political debate that should have finished fifteen years ago and while you're running the day-to-day operations of a series of dense and heavy state-owned bureaucracies. The country is crying out for a clearly stated economic agenda and it's not, in my view, that hard to get one. Central to it, though, is finishing yesterday's debates in order to define the role of government for tomorrow.

This, to me, is most naturally a Labor agenda. While the conservatives like to claim philosophical ownership of privatisation, in government they tend not to do much at all. Just as Fraser was elected without an agenda after wrecking the Whitlam government, Tony Abbott may well come to office and then spend a couple of terms simply wasting the nation's time. It's up to Labor to finish the business it started with Hawke and Keating. After fifteen years without meaningful structural economic reform, it's time to take the natural next steps.

In the absence of a clear vision of the modern role of government, people's expectations extend to the unrealistic and, sometimes, to the bizarre. In eleven years of mobile offices and Monday morning electorate email, I had a constituent demand that I shut down an abattoir next door to the house he bought sight unseen. I've been yelled at over swooping magpies – as if I was going to go out there and shoot the things myself – and I've repeatedly been told that the rising cost of living, from housing to power to petrol, is somehow the government's fault.

I have no illusions about the political popularity of a broad privatisation agenda – as a member of the most recent Queensland Labor government, it's hard not be pretty clear about the public view. But having lived through that experience, it remains my view that the only feasible way to explain to people what you're *not* going to do for them is to capture their imagination with what you are.

No, you can't give them below-cost power (and I think most Australians in their hearts know that's true), but you can redefine the nation – you can preserve a social safety net even as the nation approaches a looming demographic cliff; you can spur on the economy by developing new trade opportunities and investing in innovation; and you can make the cities and the country sustainable.

Laura Tingle's right when she says that Australians love a whinge and that right now, in the face of uncertainty, our whinge is turning angry. As a progressive, I believe there's always hope and that our current hope lies in setting an agenda for national reinvention. We need to have the guts to make unpopular decisions about what we're not going to do in order to present a clear picture of the exciting things that we are.

Rachel Nolan

George Brandis has represented Queensland in the federal Senate since 2000. He is the shadow attorney-general and served as Minister for the Arts and Sport in the Howard government.

Judith Brett is the author of three Quarterly Essays, *Exit Right*, *Relaxed and Comfortable* and *Fair Share*. Her books include the award-winning *Robert Menzies' Forgotten People*, *Ordinary People's Politics* (with Anthony Moran) and *Australian Liberals and the Moral Middle Class: From Alfred Deakin to John Howard*. She is professor of politics at La Trobe University.

Tim Flannery has published over a dozen books, including *The Future Eaters*, *The Eternal Frontier*, *The Weather Makers*, *Now or Never: A Sustainable Future for Australia?* and *Here on Earth*. He was Australian of the Year in 2007 and is presently Chief Climate Commissioner.

Mark Latham is a former leader of the Australian Labor Party and was Opposition leader from 2003 to 2005. He writes a regular column for the *Australian Financial Review* and his books include *Civilising Global Capital* and *The Latham Diaries*.

David Marr is the author of *Patrick White: A Life*, *The High Price of Heaven* and *Panic*, and co-author with Marian Wilkinson of *Dark Victory*. He has written for the *Sydney Morning Herald*, the *Age* and the *Monthly*, been editor of the *National Times*, a reporter for *Four Corners* and presenter of ABC TV's *Media Watch*. In 2010 he wrote the Quarterly Essay *Power Trip: The Political Journey of Kevin Rudd*.

Rachel Nolan was the state member for Ipswich in the Queensland parliament from 2001 to 2012. She held the portfolios of Transport, Natural Resources, Finance and the Arts as part of Anna Bligh's Labor government.

Chris Uhlmann is political editor of ABC TV's *7.30*. He has worked as the chief political correspondent for ABC radio current affairs and been political editor of *ABC News*, *The 7.30 Report* and *ABC News 24*.

Jack Waterford is editor-at-large of the *Canberra Times*, for which he has worked since 1972.

SUBSCRIBE to Quarterly Essay & SAVE nearly 40% off the cover price

Subscriptions: Receive a discount and never miss an issue. Mailed direct to your door.
☐ **1 year subscription** (4 issues): $59 within Australia incl. GST. Outside Australia $89.
☐ **2 year subscription** (8 issues): $105 within Australia incl. GST. Outside Australia $165.
* All prices include postage and handling.

Back Issues: (Prices include postage and handling.)

☐ **QE 2** ($15.95) John Birmingham *Appeasing Jakarta*
☐ **QE 4** ($15.95) Don Watson *Rabbit Syndrome*
☐ **QE 6** ($15.95) John Button *Beyond Belief*
☐ **QE 7** ($15.95) John Martinkus *Paradise Betrayed*
☐ **QE 8** ($15.95) Amanda Lohrey *Groundswell*
☐ **QE 10** ($15.95) Gideon Haigh *Bad Company*
☐ **QE 11** ($15.95) Germaine Greer *Whitefella Jump Up*
☐ **QE 12** ($15.95) David Malouf *Made in England*
☐ **QE 13** ($15.95) Robert Manne with David Corlett *Sending Them Home*
☐ **QE 14** ($15.95) Paul McGeough *Mission Impossible*
☐ **QE 15** ($15.95) Margaret Simons *Latham's World*
☐ **QE 17** ($15.95) John Hirst *"Kangaroo Court"*
☐ **QE 18** ($15.95) Gail Bell *The Worried Well*
☐ **QE 19** ($15.95) Judith Brett *Relaxed & Comfortable*
☐ **QE 20** ($15.95) John Birmingham *A Time for War*
☐ **QE 21** ($15.95) Clive Hamilton *What's Left?*
☐ **QE 22** ($15.95) Amanda Lohrey *Voting for Jesus*
☐ **QE 23** ($15.95) Inga Clendinnen *The History Question*
☐ **QE 24** ($15.95) Robyn Davidson *No Fixed Address*

☐ **QE 25** ($15.95) Peter Hartcher *Bipolar Nation*
☐ **QE 26** ($15.95) David Marr *His Master's Voice*
☐ **QE 27** ($15.95) Ian Lowe *Reaction Time*
☐ **QE 28** ($15.95) Judith Brett *Exit Right*
☐ **QE 29** ($15.95) Anne Manne *Love & Money*
☐ **QE 30** ($15.95) Paul Toohey *Last Drinks*
☐ **QE 31** ($15.95) Tim Flannery *Now or Never*
☐ **QE 32** ($15.95) Kate Jennings *American Revolution*
☐ **QE 33** ($15.95) Guy Pearse *Quarry Vision*
☐ **QE 34** ($15.95) Annabel Crabb *Stop at Nothing*
☐ **QE 36** ($15.95) Mungo MacCallum *Australian Story*
☐ **QE 37** ($15.95) Waleed Aly *What's Right?*
☐ **QE 38** ($15.95) David Marr *Power Trip*
☐ **QE 39** ($15.95) Hugh White *Power Shift*
☐ **QE 42** ($15.95) Judith Brett *Fair Share*
☐ **QE 43** ($15.95) Robert Manne *Bad News*
☐ **QE 44** ($15.95) Andrew Charlton *Man-Made World*
☐ **QE 45** ($15.95) Anna Krien *Us and Them*
☐ **QE 46** ($15.95) Laura Tingle *Great Expectations*
☐ **QE 47** ($15.95) David Marr *Political Animal*

Payment Details: I enclose a cheque/money order made out to Schwartz Media Pty Ltd. Please debit my credit card (Mastercard or Visa accepted).

Card No. ☐☐☐☐ ☐☐☐☐ ☐☐☐☐ ☐☐☐☐

Expiry date   /   **CCV**   **Amount $**

Cardholder's name   **Signature**

Name

Address

Email   **Phone**

Post or fax this form to: Quarterly Essay, Reply Paid 79448, Collingwood VIC 3066 /
Tel: (03) 9486 0288 / Fax: (03) 9486 0244 / Email: subscribe@blackincbooks.com
Subscribe online at **www.quarterlyessay.com**